# VITAL STATISTICS

## SUMMARY OF A WORKSHOP

Michael J. Siri and Daniel L. Cork, *Rapporteurs*

Committee on National Statistics

Division of Behavioral and Social Sciences and Education

**NATIONAL RESEARCH COUNCIL**
*OF THE NATIONAL ACADEMIES*

THE NATIONAL ACADEMIES PRESS
Washington, DC
**www.nap.edu**

THE NATIONAL ACADEMIES PRESS   500 Fifth Street, NW   Washington, DC 20001

NOTICE: The project that is the subject of this report was approved by the Governing Board of the National Research Council, whose members are drawn from the councils of the National Academy of Sciences, the National Academy of Engineering, and the Institute of Medicine. The members of the committee responsible for the report were chosen for their special competences and with regard for appropriate balance.

Support of the work of the Committee on National Statistics is provided by a consortium of federal agencies through a grant from the National Science Foundation (Number SES-0453930). The project that is the subject of this report was funded by allocations from the U.S. Census Bureau, the National Center for Health Statistics, and the Social Security Administration (Office of Research, Evaluation, and Statistics) to the National Science Foundation under this grant. Any opinions, findings, conclusions, or recommendations expressed in this publication are those of the author(s) and do not necessarily reflect the views of the organizations or agencies that provided support for the project.

International Standard Book Number-13: 978-0-309-14164-2
International Standard Book Number-10: 0-309-14164-8

Additional copies of this report are available from the National Academies Press, 500 Fifth Street, NW, Washington, DC 20001; (202) 334-3096; Internet, http://www.nap.edu.

Copyright 2009 by the National Academy of Sciences. All rights reserved.

Printed in the United States of America

Suggested citation: National Research Council. (2009). *Vital Statistics: Summary of a Workshop.* Michael J. Siri and Daniel L. Cork, rapporteurs. Committee on National Statistics, Division of Behavioral and Social Sciences and Education. Washington, DC: The National Academies Press.

# THE NATIONAL ACADEMIES
*Advisers to the Nation on Science, Engineering, and Medicine*

The **National Academy of Sciences** is a private, nonprofit, self-perpetuating society of distinguished scholars engaged in scientific and engineering research, dedicated to the furtherance of science and technology and to their use for the general welfare. Upon the authority of the charter granted to it by the Congress in 1863, the Academy has a mandate that requires it to advise the federal government on scientific and technical matters. Dr. Ralph J. Cicerone is president of the National Academy of Sciences.

The **National Academy of Engineering** was established in 1964, under the charter of the National Academy of Sciences, as a parallel organization of outstanding engineers. It is autonomous in its administration and in the selection of its members, sharing with the National Academy of Sciences the responsibility for advising the federal government. The National Academy of Engineering also sponsors engineering programs aimed at meeting national needs, encourages education and research, and recognizes the superior achievements of engineers. Dr. Charles M. Vest is president of the National Academy of Engineering.

The **Institute of Medicine** was established in 1970 by the National Academy of Sciences to secure the services of eminent members of appropriate professions in the examination of policy matters pertaining to the health of the public. The Institute acts under the responsibility given to the National Academy of Sciences by its congressional charter to be an adviser to the federal government and, upon its own initiative, to identify issues of medical care, research, and education. Dr. Harvey V. Fineberg is president of the Institute of Medicine.

The **National Research Council** was organized by the National Academy of Sciences in 1916 to associate the broad community of science and technology with the Academy's purposes of furthering knowledge and advising the federal government. Functioning in accordance with general policies determined by the Academy, the Council has become the principal operating agency of both the National Academy of Sciences and the National Academy of Engineering in providing services to the government, the public, and the scientific and engineering communities. The Council is administered jointly by both Academies and the Institute of Medicine. Dr. Ralph J. Cicerone and Dr. Charles M. Vest are chair and vice chair, respectively, of the National Research Council.

**www.national-academies.org**

## PLANNING COMMITTEE FOR THE WORKSHOP ON VITAL DATA FOR NATIONAL NEEDS

LOUISE RYAN *(Chair)*, Commonwealth Scientific and Industrial Research Organisation, Australia; formerly, Department of Biostatistics, Harvard University

JANET NORWOOD, Independent Consultant, Washington, DC

EDWARD PERRIN, Department of Health Services, University of Washington

SAMUEL PRESTON, Department of Sociology, University of Pennsylvania

KENNETH PREWITT, School of International and Public Affairs, Columbia University

CONSTANCE F. CITRO, *Study Director*
DANIEL L. CORK, *Senior Program Officer*
CARYN E. KUEBLER, *Research Associate (until March 2008)*
MICHAEL J. SIRI, *Program Associate*

## COMMITTEE ON NATIONAL STATISTICS
## 2008–2009

WILLIAM F. EDDY *(Chair)*, Department of Statistics, Carnegie Mellon University
KATHARINE G. ABRAHAM, Department of Economics and Joint Program in Survey Methodology, University of Maryland
ALICIA CARRIQUIRY, Department of Statistics, Iowa State University
WILLIAM DUMOUCHEL, Phase Forward, Inc., Waltham, Massachusetts
JOHN C. HALTIWANGER, Department of Economics, University of Maryland
V. JOSEPH HOTZ, Department of Economics, Duke University
KAREN KAFADAR, Department of Statistics, Indiana University, Bloomington
DOUGLAS S. MASSEY, Department of Sociology, Princeton University
SALLY MORTON, Statistics and Epidemiology, RTI International, Research Triangle Park, North Carolina
JOSEPH NEWHOUSE, Division of Health Policy Research and Education, Harvard University
SAMUEL H. PRESTON, Population Studies Center, University of Pennsylvania
HAL STERN, Department of Statistics, University of California, Irvine
ROGER TOURANGEAU, Joint Program in Survey Methodology, University of Maryland, and Survey Research Center, University of Michigan
ALAN ZASLAVSKY, Department of Health Care Policy, Harvard Medical School

CONSTANCE F. CITRO, *Director*

# Acknowledgment

This report has been reviewed in draft form by individuals chosen for their diverse perspectives and technical expertise, in accordance with procedures approved by the Report Review Committee of the National Research Council. The purpose of this independent review is to provide candid and critical comments that assist the institution in making its report as sound as possible, and to ensure that the report meets institutional standards for objectivity, evidence, and responsiveness to the study charge. The review comments and draft manuscript remain confidential to protect the integrity of the deliberative process.

The planning committee thanks the following individuals for their review of this report: Colm A. O'Muircheartaigh, Harris Graduate School of Public Policy Studies, University of Chicago; Edward B. Perrin, Department of Health Services (emeritus), University of Washington; Richard G. Rogers, Population Program and Department of Sociology, University of Colorado; and Harry M. Rosenberg, National Center for Health Statistics (retired), Bethesda, MD.

Although the reviewers listed above have provided many constructive comments and suggestions, they were not asked to endorse the content of the report, nor did they see the final draft of the report before its release. The review of this report was overseen by Linda J. Waite, Department of Sociology, University of Chicago. Appointed by the National Research Council, she was responsible for making certain that the independent examination of this report was carried out in accordance with institutional procedures and that all review comments were carefully considered. Responsibility for the final content of the report rests entirely with the authors and the institution.

# Contents

| | | |
|---|---|---|
| **1** | **Introduction** | **1** |
| | 1–A  The Workshop on Vital Data for National Needs | 3 |
| | 1–B  Successes and Challenges of the Vital Statistics Program | 4 |
| | 1–C  Report Overview | 8 |
| **2** | **Uses of Vital Statistics Data** | **9** |
| | 2–A  Uses in Health Policy and Health Research | 10 |
| |     2–A.1  Social Inequalities in Health | 10 |
| |     2–A.2  Trends in Mortality | 14 |
| |     2–A.3  Uses of Vital Statistics by the Maternal and Child Health Bureau | 17 |
| | 2–B  Population Projections and Estimates | 21 |
| |     2–B.1  Population and Fiscal Projections at the Social Security Administration | 22 |
| |     2–B.2  Population Estimates and Projections at the Census Bureau | 27 |
| |     2–B.3  Discussion | 30 |
| | 2–C  Growing and Emerging Uses: Vital Statistics and Biosurveillance | 31 |
| **3** | **The Federal-State Cooperative Relationship** | **35** |
| | 3–A  The Role of the States | 36 |
| | 3–B  Challenges and Limitations at the National Center for Health Statistics | 41 |
| | 3–C  Examples of Federal-State Cooperation in the U.S. Federal Statistical System | 42 |

|  |  | 3–C.1 The Quarterly Census of Employment and Wages | 42 |
|---|---|---|---|
|  |  | 3–C.2 The Common Core of Data | 46 |

## 4 Methodological Issues and the 2003 Revision of Standard Instruments 49
    4–A The 2003 Revisions    50
    4–B Race and Ethnicity    52
        4–B.1 Bridging Single-Race and Multiple-Race Data at NCHS    55
        4–B.2 Bridging Single-Race and Multiple-Race Data at the Census Bureau    56
    4–C Fetal Deaths and Infant Health Risk Factors    58
    4–D Mortality and Causes of Death    61

## 5 Options for a 21st Century Vital Statistics Program    65

## Appendixes    75

A The U.S. Vital Statistics System: The Role of State and Local Health Departments    *Steven Schwartz*    77

B The U.S. Vital Statistics System: A National Perspective    *National Center for Health Statistics*    87

C Workshop Agenda and Participant List    111

D 2003 Revisions, Standard Certificates of Death and Live Birth    117

References    125

# List of Figures

| | | |
|---|---|---|
| 1-1 | Flow of vital records and statistics in the United States | 2 |
| 2-1 | Population aged dependency ratio, historical and projected through 2080 | 22 |
| 2-2 | Historical and projected total fertility rates, 1915–2075 | 23 |
| 4-1 | Rates of gestational diabetes by age of mother and plurality, 12-state reporting area, 2005 | 59 |
| 4-2 | Rate of surfactant therapy by gestational age and race and Hispanic origin of mother, 12-state reporting area, 2005 | 60 |
| 4-3 | Steroids for fetal lung maturation received by the mother prior to delivery, by gestational age and race and Hispanic origin, 12-state reporting area, 2005 | 60 |

# List of Tables

4-1　Adoption of 2003 Revised Certificates and Multiple-Race Reporting for Births and Deaths, by State, 2005　　53

# List of Boxes

| | | |
|---|---|---|
| 1-1 | Successes of the U.S. Vital Statistics System | 5 |
| 2-1 | Performance and Outcome Measures for the Maternal and Child Health Bureau Block Grant Program | 18 |
| 3-1 | Surveys Comprising the Common Core of Data | 46 |
| 4-1 | Major Changes to the U.S. Standard Certificates for Vital Events, 2003 Revision | 51 |

# − 1 −

# Introduction

LITERALLY THE FIGURES OF LIFE AND DEATH, vital statistics hold an undeniable position of importance among a nation's data resources. In their basic content, the measurement of births and deaths is one of the longest-standing data collection priorities of the U.S. government, dating to at least 1850. Over the past few decades, the specific program that gathers the data has evolved into a complex cooperative program between the federal and state governments for social measurement. The vital statistics themselves are a critical national information resource for understanding public health and examining such key indicators as fertility, mortality, and causes of death, and the factors associated with them.

Vital statistics begin as individual, geographically focused vital events that are registered or certified after their occurrence. Figure 1-1 provides a basic illustration of the process by which the records of these vital events make their way into the tabulations of the vital statistics of the United States. Today, the Vital Statistics Cooperative Program (VSCP) is maintained by the National Center for Health Statistics (NCHS). Registrars in 57 vital event registration areas—the 50 states, the District of Columbia, New York City (separate from the rest of New York state), and four U.S. commonwealths and territories—collect vital event data from local officials and transmit them to NCHS. NCHS compiles those data and issues public-use data files and other products, as well as analytical reports. In the past, the VSCP also compiled records of the vital events of marriage and divorce so that it was also a critical resource for documenting changes in American family and household structures. However, budget constraints on the program in the mid-1990s, combined with declines in reporting by the registration areas, led

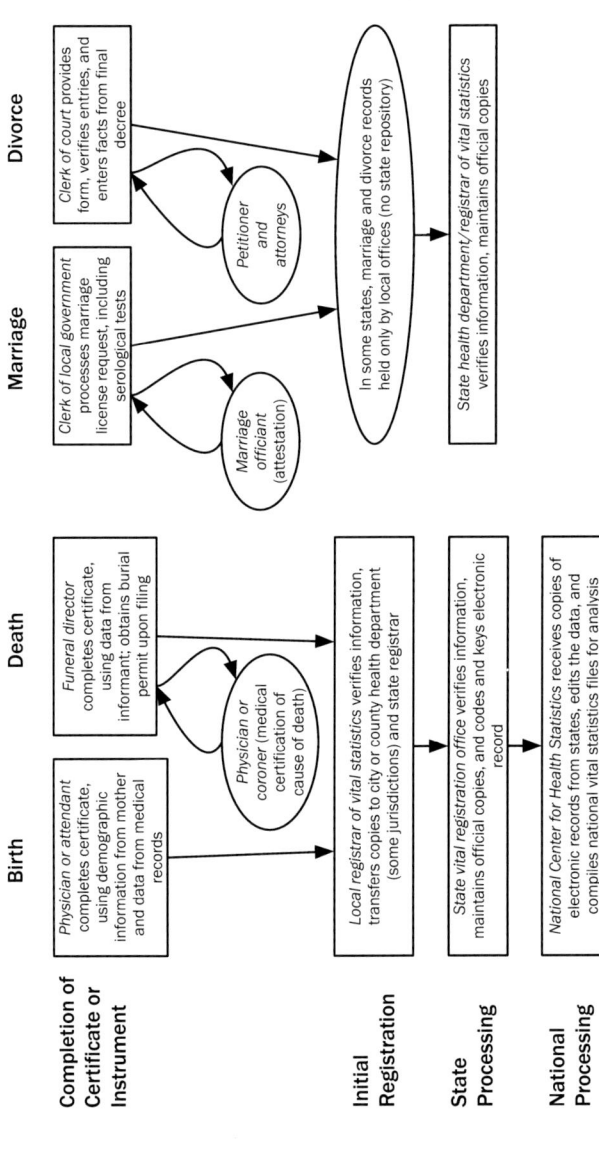

**Figure 1-1** Flow of vital records and statistics in the United States

SOURCES: Adapted from Hetzel (1997:62; reproduction of 1950 original) and National Center for Health Statistics (Appendix B, this volume:note 1). Fetal deaths are not explicitly indicated in this simplified diagram; Records of Fetal Death (bearing much of the same demographic information on characteristics of the mother as the Certificate of Live Birth) are completed and filed separately.

# INTRODUCTION

to the abandonment of the marriage and divorce series. Though marriage and divorce records are no longer compiled at the national level, the natality and mortality components of the vital statistics program have endured and continue to be essential to a wide variety of governmental and research uses.

At the workshop summarized in this report, then-Census Bureau director and former Texas state demographer Steve Murdock marveled at the degree to which the vital statistics on birth and death are taken for granted. He observed that these statistics have grown to be sufficiently critical to so many processes and assessments of the nation's well-being that it is assumed that they always have and always will exist. Yet that is the paradox of vital statistics: data on births and deaths seem so fundamental and—at first glance—so simple a metric of social conditions that their existence is assumed to be automatic and their collection is assumed to be easy. A history of vital records collection commissioned in 1950 (Hetzel, 1997:43) captured this theme well:

> Most people take vital statistics for granted, assuming that any statistics they need should be freely available as part of today's culture. . . . Death rates are among the typical vital statistics that most people assume we have always had available and, without much effort, will continue to have. The real story is quite different: national statistics of deaths and births were achieved only within the present generation, after two centuries of intermittent struggle and building.

The U.S. vital statistics system relies on the original information reported (and the consistency of that reporting) by myriad physicians, new parents, and funeral directors; channeled through state and local information systems of widely varying levels of sophistication and automation; and coordinated and processed by a federal statistical agency that has experienced relatively flat funding for many years. The challenges facing the vital statistics system and the continuing importance of the resulting data make it an important topic for periodic examination, assessing both current and emerging uses of the data and considering the methodological and organizational features of compiling vital data.

## 1–A THE WORKSHOP ON VITAL DATA FOR NATIONAL NEEDS

Pursuant to its charge to improve the statistical information and methods on which public policy decisions are made, the Committee on National Statistics (CNSTAT) of the National Academies convened a Workshop on Vital Data for National Needs on April 30, 2008. The workshop was designed to consider "the critical importance of adequate vital statistics for the statistical, research, and policy communities" and "improvements that are needed at NCHS's vital statistics program."

CNSTAT organized the workshop with support from its core sponsors, as well as additional support from the Census Bureau, the Office of Research, Evaluation, and Statistics of the Social Security Administration, and NCHS. The workshop drew nearly 100 participants, including the invited speakers and discussants. Prior to the workshop, the workshop's planning committee asked that two background papers on two basic perspectives on the vital statistics be prepared in order to inform the discussion. Steven Schwartz (New York City Department of Health and Mental Hygiene) prepared a paper on the role of the states and vital registration jurisdictions and NCHS staff prepared one on the national-level, administrative perspective; these background papers are presented in Appendixes A and B of this summary. Appendix C adds information on the workshop agenda and lists the workshop participants.

## 1–B SUCCESSES AND CHALLENGES OF THE VITAL STATISTICS PROGRAM

Workshop remarks by Harry Rosenberg (NCHS, retired) identified 22 major successes in the current vital statistics program; the full set of these successes is described in Box 1-1 and his comments provide a useful orientation to the range of topics touched on by the workshop.

Rosenberg elaborated on four particularly notable successes, each of which he said represents highly effective collaboration between the states and the federal government.

1. *Production of the annual vital statistics files:* Rosenberg argued that the greatest success of the VSCP is its basic regular product—statistical files covering 6–7 million records, including about 4 million live births and 2.5 million deaths. He said that the complexity of producing these statistical files cannot be overstated; production requires constant interaction between the states and NCHS in terms of receipt and control, intensive processing and quality evaluation, and the preparation of tables and public-use data files for analysis.

2. *Revisions of the U.S. standard certificates:* Rosenberg observed that it is important to acknowledge the periodic revisions of the standard birth and death certificates as an achievement of the program (see Chapter 4 for a detailed discussion and Appendix D for reproductions of the current standard instruments). Through these periodic updates, both administrative and research ends are achieved: they permit national compilation to keep abreast of the changing legal and administrative environment in the reporting areas while improving the data content of the final files to be of greater use to public health officials and researchers.

**Box 1-1**  Successes of the U.S. Vital Statistics System

- *Vital Statistics of the United States Data Files*—production of annual statistical data files based on about 6–7 million records of births and deaths
- *Model Legislation*—1907 template for state legislation on birth and death registration
- *Model Vital Statistics Act*—fuller version of model legislation that expanded coverage of system to include marriage and divorce records; first suggested to states in 1942 and revised in 1959 and 1977
- *Completing the National Vital Registration Areas for Births and Deaths*—as of 1933, all 48 states and the District of Columbia had adopted laws consistent with the model legislation, adopted the suggested birth and death certificates, and reported 90 percent (or greater) total registration of events
- *Tests of Birth Registration Completeness*—series of three experiments (1940, 1950, 1964–1968) conducted by the Census Bureau to verify the completeness of reporting; 1964–1968 study suggested over 99 percent registration of births
- *Query Programs*—development of manuals and training materials to assist source reporters (e.g., physicians coding causes of death) in consistently completing vital record data items
- *Current Mortality Sample (discontinued)*—beginning in the 1940s, state registration areas directly forwarded a 10-percent sample of incoming death records to the national vital statistics office, thus enabling publication of national estimates of causes of death with only a 4-month lag after month of occurrence; discontinued around 1995 because of resource constraints
- *Classifying Causes of Death According to International Standards*—World Health Organization's International Classification of Diseases (ICD) (and subsequent revisions) adopted as coding standard for U.S. national vital statistics since 1900
- *Comparability Studies*—tests of consistency of classification and coding of cause-of-death data after implementing new ICD revisions, last done after current ICD revision (ICD-10) implemented in 1999
- *Ranking Leading Causes of Death*—1951 standard developed by national and state vital statistics offices for producing ranked lists of causes of death, separately for infants and noninfants
- *Mortality Medical Data System*—set of software programs, originally developed in the late 1960s, to simplify cause-of-death coding and consistently resolve multiple cause-of-death codes; the programs have been adopted by other countries as well as the states, and recent revisions have worked toward a goal of permitting natural language entry of death causes rather than numeric codes
- *Race and Ethnicity Data (see Chapter 4)*
- *Fetal Death Reporting*—data on stillbirths have been collected since 1939, and World Health Organization standards for defining fetal death were adopted in 1950
- *Abortion Reporting (discontinued)*—From the 1970s through 1993, NCHS and the states worked on a reporting system for induced terminations of pregnancies; the system was ended in 1993 because of resource constraints

*(continued)*

> **Box 1-1**  (continued)
>
> - *Follow-back Surveys*—periodic surveys to collect additional information on samples of birth and death records, conducted as early as the mid-1950s; effectively discontinued because of resource constraints (none conducted since 1993)
> - *Training*—training sessions on medical coding and specific methodological techniques (e.g., using the Mortality Medical Data System) were conducted by NCHS beginning in 1983, but have been reduced greatly in number in recent years
> - *Mortality Workshops*—wide-ranging practitioner workshops on improving cause-of-death data convened by NCHS in 1989 and 1991
> - *Electronic Registration of Vital Events (see Appendixes A and B)*
> - *Electronic Microdata Sets*—public-use data files of vital statistics on birth and death, with measures to protect the confidentiality of individuals, have been available in various formats since 1968: data tapes, CD-ROM, and most recently via the Internet
> - *Revisions of the U.S. Standard Certificates (see text and Chapter 4)*
> - *The Linked File of Infant Deaths and Live Births (see text)*
> - *The National Death Index (see text)*
>
> SOURCE: Adapted from workshop presentation by Rosenberg and his follow-up paper (Rosenberg, 2008).

3. *Creation and development of the linked file of infant deaths and live births:* Rosenberg noted that infant mortality is one of the most widely used measures of overall health of a community. He said that, for this reason, the creation of a national linked file of live births and infant deaths has been a remarkably useful tool for understanding the medical circumstances and causes of death of infants and informing possible interventions to curb specific infant mortality types.

4. *Creation and development of the National Death Index:* Prior to 1979, researchers who wanted to conduct epidemiological studies on cause of death (particularly following up on previous studies using human subjects) had to contact every registration area separately and make arrangements for a death records check, purchasing death certificates that matched their study subjects, and coding the relevant information. Given the cumbersome nature of this process, the 1982 creation and continued updating of the National Death Index—a compilation of over 62 million death records for 1979–2005 (as of September 2007)—has been an invaluable resource for research. Rosenberg estimated that the National Death Index has assisted 1,500 research projects by performing about 4,300 searches. Among the noteworthy projects to make use of the index data are drug surveillance studies by pharmaceutical companies, evaluations of the cancer registries of

INTRODUCTION 7

the National Cancer Institute and the Centers for Disease Control and Prevention, and studies of post-employment death due to exposure to hazardous substances conducted by large oil and chemical companies.

Yet the current vital statistics cooperative program faces significant challenges; indeed, Rosenberg's tally of some of the key successes was accompanied by his notation of several shortcomings:

- *Struggle for timeliness in data production:* As a cooperative partnership, the timeliness and quality of vital statistics rise and fall with the input of contributing registration areas. Physically, compiled files for the nation as a whole cannot be put together and released until the last state or registration area submits its data. Rosenberg also noted that the complexity of the data adds to the lag time between the end of a calendar year and when birth and death data for that year become available. At the time of the 2008 workshop, the most recent available vital statistics data covered births and deaths in 2005; the lag time between close of the data year and publication of the final data is 24–25 months. Rosenberg cited a survey of vital records participation by Friedman (2007), observing that this lag time has varied widely between 1985 and 2004, from less than 2 years to as many as 4 years.

- *Difficulty in achieving adoption of 2003 certificate revision:* The most recent revision of the standard certificates of live birth and of death have been slow to win acceptance by the registration areas (see Chapter 4). This slow adoption has been particularly problematic because the revision implemented new standards for permitting reporting of multiple-race categories; until full compliance is achieved, the "national" vital statistics data are a patchwork of different reporting formats and styles for a critical data item.

- *Discontinuation of national collection of some vital records and downgrading of some quality assurance methods:* There have been major casualties of data streams within the national vital statistics collections, which Rosenberg attributed principally to the inability to secure adequate and sustained funding for the system. The most prominent of these casualties is data on marriage and divorce; marriage and divorce records do continue to be developed at the state and local level but national-level collection and compilation was discontinued almost 20 years ago because of budget concerns. A portion of the vital statistics program also briefly collected national-level data on terminated pregnancies and the circumstances—data that, objectively compiled, would inform the ongoing national debate on abortion—but that system was also discontinued: if it were still in operation, Rosenberg said that it would add over 1 million additional records to annual vital statistics. Other reductions have been more subtle but are still very consequential. Rosenberg said that NCHS has stopped coding occupation and industry of decedents, which can be important markers of both socioeconomic status and possible deaths due to workplace characteristics. The current mortality sample (meant as a quicker system for surveillance of death types) and the natality and mortality followback surveys (used for quality assurance) have been dropped (or effectively discontinued). Budget constraints have also led to reductions in NCHS-provided training courses for vital record collectors at the state and local level.

## 1–C  REPORT OVERVIEW

This workshop summary largely follows the topic blocks that were used in scheduling the workshop, though some rearrangement has been made when that seemed logical. Following this introduction, Chapter 2 briefly describes the current uses of vital statistics as presented at the workshop, particularly their use in deriving population estimates and various projections. Chapter 2 also discusses the emerging field of public health surveillance and the possible roles for vital statistics in that framework. In Chapter 3 we turn to the structure of the existing VSCP, from both the state or registration area perspective and NCHS's perspective as the national-level coordinator and primary funder of the system. The workshop featured selected case studies of analogous partnership systems in the federal statistical system, and those are briefly recounted in the chapter. Chapter 4 considers methodological issues and, in particular, those raised by the 2003 revision of the standard birth and death certificates, which includes a new format for race and Hispanic origin data and preliminary findings from new public health data items included on the certificates. Finally, Chapter 5 summarizes the concluding session of the workshop, which identified different possible visions for the vital statistics program and featured a roundtable set of reactions from a discussant panel.

# – 2 –

# Uses of Vital Statistics Data

FROM THE OUTSET, an intended purpose of the Workshop on Vital Data for National Needs was to provide information on the range of uses of the current vital statistics data and to suggest important uses on the immediate horizon. Given the tight time constraints of a 1-day session, the workshop zeroed in on two major classes of current uses: public health research and the development of population estimates and projections.

With regard to health policy and health research, summarized in Section 2–A, workshop presentations focused on two major demographic phenomena of long-standing interest: disparities or inequities in health across different racial and ethnic subgroups and gender differences in mortality. This session of the workshop also contrasted these academic perspectives on the uses of vital statistics data with the use of the data for program and planning purposes by the Maternal and Child Health Bureau (MCHB) in the U.S. Department of Health and Human Services. In Section 2–B, we summarize workshop presentations and discussion on the development of population projections and estimates by the Census Bureau and the Social Security Administration; in the latter case, the decades-long projections of population composition based on vital statistics play a key role in the major policy debates on the long-run viability of Social Security entitlements. In terms of future directions, Section 2–C summarizes the workshop's session that focused on the emerging field of biosurveillance—monitoring of disease and mortality with fine spatial and temporal precision in order to rapidly detect major disease outbreaks or, perhaps, terrorist attacks using biological agents.

## 2–A  USES IN HEALTH POLICY AND HEALTH RESEARCH

### 2–A.1  Social Inequalities in Health

Nancy Krieger (Harvard School of Public Health) spoke on the use of vital statistics and related data to monitor health inequities in the United States—studies of trends in health and health care as they are related to socioeconomic position, ethnicity, and gender. Her remarks summarized findings from her Public Health Disparities Geocoding Project. Detailed information on the project and related publications are available online at http://www.hsph.harvard.edu/thegeocodingproject (April 2009).

The project's objective is to augment data in public health surveillance systems, including the birth and death certificate data, with additional socioeconomic covariate information; the resulting constructs are termed area-based socioeconomic measures (ABSMs). The methodology links geocoded vital statistics and U.S. census data at the block group, census tract, and ZIP code tabulation area levels of geography. Ultimately, the intended goal is to develop a valid, robust, easy-to-construct, and easy-to-interpret ABSM that can be readily used by any U.S. state health department or health researcher for public health monitoring and for studying any health outcome from birth to death for any age, gender, or racial or ethnic group. The project started in 1998, making use of data from the Massachusetts Department of Public Health and the Rhode Island Department of Health; the data were for a set of years centered around the 1990 census, and the socioeconomic data in the ABSMs made use of information from that census.

To test robustly whether choice of ABSM and geographic level matters, Krieger said that she focused on a wide variety of health outcomes, including mortality (all cause and cause specific), birth (specifically, low birth weight) and also cancer incidence (all sites and site specific), childhood lead poisoning, sexually transmitted infections, tuberculosis, and nonfatal weapons-related injuries. Each outcome was analyzed in relation to 19 different ABSMs, capturing diverse aspects of socioeconomic position. Eleven of the measures were single-variable measures (e.g., percent working class, percent crowded household) and eight were composites (e.g., deprivation indices developed in previous research). Analyses were performed for the total population and also stratified by race, ethnicity, and gender.

Krieger summarized four key findings from the geocoding project. First, measures of economic deprivation were most sensitive to the expected socioeconomic gradients in health. Second, census-tract-level analyses yielded the most consistent results, with maximal geocoding, compared to the block group and ZIP code data. Third, these findings held for separate analyses conducted for white, black, and Hispanic men and women; they also held for those outcomes that could be meaningfully analyzed among the smaller Asian, Pacific Islander, and American Indian populations. Fourth, the single-variable measure of percentage of persons below poverty performed as well as more complex composite measures of economic deprivation, including the Townsend index.[1] The research suggested that socioeconomic inequalities in health are best monitored with a census-tract poverty measure; Krieger said that one advantage of this approach is that the measure can be applied to all persons, regardless of age, gender, current individual-level educational status, or current employment status.

Krieger presented socioeconomic gradients for several health outcome measures to illustrate that the technique provides a way for routine documentation and monitoring of trends using existing vital statistics and public health surveillance data. Specifically, her graphic displays divided census tracts into categories based on percentage of the population below the poverty level (e.g., less than 5 percent, 20 percent or greater). The figures suggested clear poverty gradients in terms of

- *low birth weight*, the risk of which was two times higher among births occurring in the most versus least impoverished tracts, that is, 7.5 percent versus 3.6 percent;

- *children with elevated lead levels*, with a seven-fold excess among those living in the most versus least impoverished census tracts (33 versus 5 percent);

- *syphilis*, with excess risk for the most impoverished tracts being 17 times higher than for the least impoverished tracts;

- *cervical cancer*, the incidence of which was twice as high for the most impoverished areas (18 versus 9 per 100,000 population);

- *nonfatal gunshot injury*, with an 11-fold increase (22 versus 2 per 100,000 population); and

- *heart disease mortality*, with a 1.4-fold excess risk found, resulting in an excess of nearly 100 deaths per 100,000 population.

Moving to analysis of racial, ethnic, and gender health disparities, Krieger presented 1989–1991 data on premature mortality (death before age 65). As context, the data indicated that fully half of the black and Hispanic populations lived in census tracts with 20 percent or more of the population below the poverty level whereas, by contrast, almost 50 percent of white men and women live in census tracts with less than 5 percent below poverty. Against this demographic backdrop, the researchers found evidence

---

[1] The Townsend index (Townsend, 1987; Townsend et al., 1988) is a composite index score based on four area-based census measures: percentage of households with no car, percentage of households not owner-occupied, percentage of persons unemployed, and percentage of households overcrowded.

of marked socioeconomic disparities in premature mortality, with the estimated relative risks ranging from 1.6 to 2.8. Within each economic stratum, an excess of premature mortality remained apparent among the black population. Looking a decade later (1999–2001 data), the same trends persisted: for white non-Hispanic, black, and Hispanic men, higher levels of census-tract poverty were associated with an elevated risk of dying prematurely, with black and Hispanic populations most likely to live in the most impoverished census tracts.

Krieger noted similar trends in heart disease mortality data from the period 2000–2005 for Massachusetts. Without disaggregation by poverty level, age-standardized heart disease rates among men and women show a basic distinction, with blacks at higher risk than whites. However, stratifying by census-tract poverty level shows more complex gradients: the poorest census tracts have consistently higher risk levels than the least poor, with particularly pronounced gaps for white and black men living in the poorer census tracts. Similar findings follow from an analysis using 2004–2005 Massachusetts birth outcome data involving low birth weight and smoking during pregnancy. The analysis suggests that racial and ethnic disparities again exist within each socioeconomic stratum, with blacks doing worse for low birth weight and whites doing worse in terms of smoking. There are also marked socioeconomic gradients within each racial or ethnic group. Analysis of these data is ongoing, with the final report slated to include data on prenatal care, breast feeding, caesarian sections, preterm deliveries, and infant mortality.

Krieger said that sharing data, methods, and publications on the project website is an important part of the project's goal to enhance the data reported by U.S. state health departments. Project researchers have conducted training sessions of personnel at health departments, and the techniques have been used in special reports issued by several states, including Washington and Maryland. The intent for the project is to expand the state health departments' use of geographic analysis in analyzing vital statistics.

Krieger noted recent work done in collaboration with the Boston Public Health Commission and the Massachusetts Department of Public Health to extend the work to city-defined neighborhoods and to portray socioeconomic and health data on a consistent set of maps. The system developed by the researchers concentrates on premature mortality as the outcome measure; the analysis system is built on modeling premature mortality as a function of fixed and random effects, allowing for statistical smoothing in the estimation of small-area rates, estimation of variance at each of the specified levels, and adjustment for multiple covariates. A particularly interesting finding from this work was based on mapping the population-based proportion of premature deaths that would *not* have occurred if residents in every census tract enjoyed the same age-specific mortality rates as residents of the least impoverished tracts. Krieger said that this proportion exceeded 20 percent for 8 of Boston's 60 neighborhoods and 68 percent of the city's census tracts. In two of Boston's poorest neighborhoods—Roxbury and North Dorchester—the high excess fractions suggest that, in more than half their census tracts, some 25–30 of every 100 deaths among people under age 75 would not have happened if people in those neighborhoods had, at each and every age, the same lower risk of dying as people in the richer areas.

Recently, the project considered U.S. national trends and inequities in premature mortality from 1960 through 2002. County-level mortality data from the National Center for Health Statistics (NCHS) were linked to county-level population and median family income data from the Census Bureau. These data were used to calculate and compare premature mortality and infant death rates by county income quintile for the entire study period. The study found that, even as premature mortality declined in all county income quintiles, the gap between the lowest and highest income quintiles persisted over the entire period and it was relatively greatest for premature mortality in 2000. The greatest progress in reducing these income gaps occurred between 1965 and 1980, especially for populations of color; thereafter, the health inequities widened. The same pattern held for infant deaths. The researchers also used an approach similar to that in the Boston neighborhood study, considering excess premature deaths that would not have occurred if the rates in the least impoverished areas were the same as those for the most impoverished areas. Under these assumptions, Krieger said that the research showed that, had everyone experienced the same yearly age-specific mortality rates as whites in the highest-income-county quintile between 1960 and 2002, 14 percent of white and 30 percent of nonwhite premature deaths would have been averted.

Going forward, a challenge will be working with a new data source. Unlike the 1990 and 2000 censuses, the 2010 decennial census will not include a long-form sample that obtains additional social and demographic information (including questions used to calculate census-tract-level poverty estimates). Instead, that information is now covered in the Census Bureau's American Community Survey (ACS). The ACS provides the same data items as the old long form but, because it is collected on a continuous basis (spreading the sample out over several years), the data are in a new format: rolling averages based on 1, 3, or—for small areas such as tracts—5 years of data. Krieger indicated that project researchers are beginning work to explore how best to develop the tract-level characteristics based on ACS data.

Krieger concluded that vital statistics are critical for understanding current and changing U.S. patterns of health and health inequities and the story they tell is compelling. Krieger noted that some of these themes were expressed in a 2008 PBS documentary, *Unnatural Causes: Is Inequality Making Us Sick?* The basic data of vital events are core to these public education efforts, because they alone can reveal whether population health and health inequities are getting better or worse.

## 2–A.2 Trends in Mortality

Richard Rogers (University of Colorado) began his remarks by commenting that there was a period, in the 1970s and 1980s, when it was generally thought that the important questions related to the study of mortality had already been asked and that the set of factors influencing mortality were well understood. Thirty years of subsequent research demonstrates that the study of mortality remains one of critical importance to understanding health in the United States. As illustrated by Krieger's presentation, widespread disparities in health and longevity are one important reason for further study of mortality trends. Rogers said that mortality studies are also important because mortality affects a variety of different, broader factors, including social relationships and social institutions; it can have a profound influence on individuals, on families, on communities. It is important to social policies and population forecasting; in thinking of health care financing in the long run, mortality studies are of central importance for administration of Social Security and Medicare.

International comparisons are a major emerging motivating factor for studies of mortality. Specifically, Rogers noted a study by Banks et al. (2006) that found a fairly large disparity between the American and English populations. Rogers summarized the study as having two major findings: first, that prevalence rates for disease were generally higher for Americans than for the English and, second, that the socioeconomic health status gradient is a real construct and is evident in both countries. Generally, Rogers said that the fact that the United States is not at the top of the world in terms of life expectancy—there are at least 22 other countries with longer life expectancies—is a basic motivational factor for further study of basic questions: Why are Americans sick and why does the U.S. life expectancy lag behind that of other countries?

Because of time constraints, Rogers centered his remarks on sex differences in life expectancy. The data used in his research include mortality data from the vital statistics, particularly a linked mortality file combining records from the National Death Index with survey data from NCHS's National Health and Nutrition Examination Survey (NHANES). The research also uses data from NCHS's National Health Interview Survey.

Analysis of estimated life expectancy at birth from 1900 to the present shows generally increasing life expectancies for both males and females, though expectancies for males are consistently lower than those for females. Shifts in the data show the effects of infection for several periods, especially the influenza epidemic in 1918. After greater control for infectious diseases, mortality becomes less volatile from the 1940s onward. However, the data also show a slow convergence of the male and female trend lines as the gender difference in life expectancy narrows. After peaking at a 7.8-year difference in 1975 (Arias, 2007), the difference between men and women in estimated life expectancy has steadily declined: by 2005, the gap was 5.2 years.

Rogers noted that many studies have looked at the differences around the 1978 peak, but fewer studies have examined the motivating factors for the subsequent decline in the gap. He briefly suggested a range of possible factors that contribute to sex differences in mortality: biological factors, health behaviors (smoking, drinking, unsafe driving, exercise), environmental risks, social relations (marriage, family composition), and socioeconomic status (education, employment, income, poverty). Rogers suggested that some as-yet-underresearched possibilities include composite measures that may be difficult to pick up in national data sets. One is addressing the concepts and assumptions of "masculinity" and "femininity"—for instance, the extent to which "masculine" traits of a high pain threshold, reluctance to seek medical help (absent a life-threatening condition), and failure to get regular health checkups affect health outcomes. The differential life expectancy by sex still shows up when mortality rates are disaggregated by age. The biggest age gap between males and females manifests itself in late teens and early adolescence, what Rogers said has been described as the "accident peak" or "testosterone spike."

Cigarette smoking patterns are one variable that seems to be a central contributor to sex differences in mortality, but those patterns have changed over time. Historically, males have tended to smoke in higher proportions than females—about 53 percent of adult men smoked in 1955, compared with 25 percent of adult women. However, over time, rates of smoking have decreased for both sex groups although females have drawn closer to males (an estimated 24 percent of adult men reported smoking in 2004, compared with 18 percent of adult women). Rogers cited previous research in concluding that smoking contributes to some of the sex differences in mortality and life expectancy. Retherford (1972) attributed 47 percent of the sex gap in life expectancy in 1972 to cigarette smoking; Rogers' own work with colleagues (Hummer et al., 1998; Rogers et al., 2000) suggests that smoking contributed to about 25 percent of the gap as measured in 1990–1995. These estimates are consistent with an overall decline in smoking and a convergence between males and females in their smoking patterns.

Rogers summarized work with hazard ratios derived from NHANES data for 1988–2000. Though the original intent of the work was to try to explain away of the sex difference in mortality, the results actually suggest more explanations for a widening of the gap than a narrowing. Relative to males, females in this period had less education, had lower incomes, and were less likely to be employed—that is, they were disadvantaged on a number of socioeconomic status measures. Once these factors are controlled, the hazard ratio expands and the gap in mortality widens. Controlling for marital status also widens the gap; this finding can be explained by males' tendency to marry younger women but die at earlier ages, meaning that females end up living longer in a widowed status. Rogers also noted that religious attendance has some influence on the sex differential (reducing the gap), because females are more religious and attend services more frequently. Physical activity tends to widen the gap, as does disability (as measured by a question on difficulty in walking).

Examining causes of death—looking at sex differences in mortality associated with specific causes rather than overall—provides additional insight about the sex gap. Rogers noted that the gap is particularly wide for deaths due to circulatory disease and cardiovascular disease, while cerebrovascular diseases have less role in explaining the differences between males and females. The significant sex differences in terms of deaths due to cancer are mostly a result of cigarette smoking; the major difference (higher rates of lung cancer mortality among men) disappears when smoking is considered. Respiratory diseases do not have a significant difference between the genders, but deaths due to external causes (accidents, homicides, and suicides) do; because of small sample sizes, these effects are hard to examine in detail.

Rogers concluded that part of his results are based on specific periods, specific durations, and specific follow-up time periods. Period effects are important—researchers get different results in explaining sex differences in longevity and mortality in the 2000s than were estimated in the 1970s and 1980s. Still, it is important to think about other covariates and, specifically, what other covariates might be important that are not regularly collected in current national surveys and national data sets. Such covariates could include geographic information; they could include better measures of religiosity or religious attendance; and they could also include such factors as altruism, genetics, biology, stress, and refined quantification of socioeconomic status.

In discussion, Rogers noted that the existing interview data from the National Health Interview Survey and NHANES are generally restricted to the noninstitutional population: understanding the degree to which these survey measures are conservative estimates (because they exclude major segments of older persons in nursing facilities and younger persons in correctional facilities) is an important consideration for future research. It was also noted that deaths of U.S. citizens overseas—and, particularly, military deaths—are not included in standard vital statistics (and, hence, not in general assessments of health inequities that use those data). Rogers concluded that health disparities are important and reducing them is a critical national objective for the United States; he said that we need more information to more fully understand some of the differences, by sex, by age, by race and ethnicity, and by socioeconomic status.

### 2–A.3 Uses of Vital Statistics by the Maternal and Child Health Bureau

Peter van Dyck (MCHB, Health Resources and Services Administration, U.S. Department of Health and Human Services) described the various ways in which vital statistics are used by MCHB:

- as the basis for both assessing eligibility for and monitoring performance of targeted public health grants;
- as input to regular publications and policy standards; and
- as a way of evaluating an agency's progress toward general objectives.

He also commented on MCHB's role in issuing grants to help states reengineer their vital statistics and child health information systems.

Pursuant to Title V of the Social Security Act of 1935, the MCHB is responsible for providing a variety of grant and coordination services. The bureau's responsibilities make it the oldest continuing health program related to mothers and children in the nation. Each year, MCHB administers about $1 billion in grants, most of which—about $600 million—is provided as block grant allocations to the states and territories. The block grant funds are allocated using a formula based on a state's percentage of children living in poverty as a share of the national total; the funds support the operation of state-level maternal and child health offices and programs. Van Dyck said that the states are required to provide matching funds (at least $3 in state funds for every $4 in federal funds), which the states usually generate by billing Medicaid or private insurance for the services they deliver to maternal and child health clients. Some counties also provide funds or staff support. In this way, the $600 million in federal money for maternal and child health grants is leveraged to yield a total effort of $5 billion to 6 billion.

To qualify for and obtain the MCHB Title V block grants, state applicants must annually report on a series of 18 specific performance measures; see Box 2-1. Van Dyck noted that vital statistics are essential to this performance and evaluation effort, because several of the performance measures are obtained directly from vital records data (as indicated in italics in the box). State grantees are also directed to provide regular information on a set of six national performance outcome measures, also shown in the box; all of these are directly computed from vital statistics.

The Title V block grant program also makes use of a set of "health system capacity indicators" (HSCIs) and "health status indicators" (HSIs) in program evaluation, several of which are keyed directly to vital statistics:

- *HSCI #04:* Percentage of women ages 15–44 with a live birth during the reporting year for whom the ratio of observed to expected prenatal

**Box 2-1** Performance and Outcome Measures for the Maternal and Child Health Bureau Block Grant Program

**Performance Measures**
1. Percent of screen positive newborns who received timely follow-up to definitive diagnosis and clinical management for condition(s) mandated by their state-sponsored newborn screening programs
2. Percent of children with special health care needs age 0–18 whose families partner in decision making at all levels and are satisfied with the services they receive
3. Percent of children with special health care needs age 0–18 who receive coordinated, ongoing, comprehensive care within a medical home
4. Percent of children with special health care needs age 0–18 whose families have adequate private and/or public insurance to pay for the services they need
5. Percent of children with special health care needs age 0–18 whose families report the community-based service systems are organized so they can use them easily
6. Percent of youth with special health care needs who received the services necessary to make transitions to all aspects of adult life, including adult health care, work, and independence
7. Percent of 19–35-month olds who have received full schedule of age appropriate immunizations against measles, mumps, rubella, polio, diphtheria, tetanus, pertussis, haemophilus influenza, and hepatitis B
8. *Rate of birth (per 1,000) for teenagers ages 15–17 years*
9. Percent of third-grade children who have received protective sealants on at least one permanent molar tooth
10. *Rate of deaths to children ages 14 years and younger caused by motor vehicle crashes per 100,000 children*
11. Percent of mothers who breast-feed their infants at 6 months of age
12. Percent of newborns who have been screened for hearing before hospital discharge
13. Percent of children without health insurance
14. Percent of children, ages 2–5 years, receiving WIC services that have a Body Mass Index (BMI) at or above the 85th percentile
15. *Percent of women who smoke in the last 3 months of pregnancy*
16. *Rate (per 100,000) of suicide deaths among youths 15–19*
17. *Percent of very-low-birth-weight infants delivered at facilities for high-risk deliveries and neonates*
18. *Percent of infants born to pregnant women receiving prenatal care beginning in the first trimester*

**Outcome Measures**
1. *Infant mortality rate per 1,000 live births*
2. *Ratio of the black infant mortality rate to the white infant mortality rate*
3. *Neonatal mortality rate per 1,000 live births*
4. *Postneonatal mortality rate per 1,000 live births*
5. *Perinatal mortality rate per 1,000 live births plus fetal deaths*
6. *Child death rate per 100,000 children ages 1–14*

NOTE: Italics indicate that the measure is derived from vital statistics data.

SOURCE: Workshop presentation by Van Dyck; http://mchb.hrsa.gov/training/performance_measures.asp (April 2009).

visits is greater than or equal to 80 percent on the Kotelchuck Index (which is related to the mother's age at time of prenatal care entrance and the birth weight of the baby if the baby is born early)
- *HSCI #05:* Comparison of infant deaths between Medicaid and non-Medicaid recipients, using information associated with prenatal care, low birth weight, and infant mortality. (Van Dyck added that the MCHB's website, which posts these indicators for all grant recipients, is the only ongoing data site that provides the rate of infant deaths for Medicaid clients compared with the infant deaths for non-Medicaid clients.)
- *HSCI #09A and B:* Self-scores by the states on their data capacity for implementing four types of data linkages:
  - annual linkage of infant birth and infant death certificates
  - annual linkage of birth certificates and Medicaid eligibility or paid claims files
  - annual linkage of birth certificates and WIC eligibility files
  - annual linkage of birth certificates and newborn screening files
- *HSI #01A:* Percent of live births weighing less than 2,500 grams
- *HSI #01B:* Percent of live singleton births weighing less than 2,500 grams
- *HSI #02A:* Percent of live births weighing less than 1,500 grams
- *HSI #02B:* Percent of live singleton births weighing less than 1,500 grams
- *HSI #03A:* Death rate per 100,000 due to unintentional injuries among children ages 14 years and younger
- *HSI #03B:* Death rate per 100,000 for unintentional injuries among children ages 14 years and younger due to motor vehicle crashes
- *HSI #03C:* Death rate per 100,000 for unintentional injuries for youth ages 15 through 24 years due to motor vehicle crashes

MCHB also administers the $100+ million, county-based Healthy Start program, which is intended to reduce infant mortality rates in vulnerable or poor communities. The program is intended to facilitate service delivery in selected areas, including easing access to prenatal health care and promoting positive prenatal health behaviors. As with the block grants, Healthy Start administrators depend on vital statistics—in this case, detailed disaggregation of infant mortality rates—to target program activities and evaluate progress. Looking at county-level plots of infant mortality rates is a particularly important diagnostic tool for Healthy Start, allowing MCHB to pinpoint areas in the nation that might be eligible to apply for grants (grant funds are only available to areas with high rates of infant mortality). Birth certificate data on low birth weight are analyzed in the same manner, and new covariate information on prenatal health care and behaviors on the most recent version of the birth certificate (see Chapter 4) are emerging as important assessment tools. The Healthy Start program periodically issues reports to Congress on the progress of the funding, and calculations based on vital statistics are critical in justifying the program to congressional sponsors.

Van Dyck also indicated that MCHB is responsible for compiling data related to its work for other government publications, including the bureau's own regular publications *Child Health USA* and *Women's Health USA*. Most notably, MCHB was responsible for developing the "Maternal and Child Health" chapter of *Healthy People 2010* (U.S. Department of Health and Human Services, 2000), an extensive set of public health measures and targets for the nation as a whole.

MCHB's use of vital statistics to describe progress toward *Healthy People 2010* and other objectives is intended to serve as a diagnostic of behavior by the general public. However, van Dyck also observed that the measures also serve as the benchmark by which MCHB and its programs are evaluated by the Department of Health and Human Services and Congress. On the basis of the vital statistics indicators, van Dyck said that MCHB is held accountable by its departmental superiors. The indicators can become the basis for specific MCHB programs—for instance, efforts to achieve reductions in sudden infant death syndrome, to provide earlier recognition of autism, or to increase daily intake of folic acid (to reduce the incidence of spina bifida and other disorders). More generally, a major part of MCHB's overall mission is the reduction of health disparities, and MCHB is beginning to use vital statistics (infant mortality rates) and measures of socioeconomic deprivation (akin to Krieger's work, presented at the workshop) to assess the bureau's progress toward that goal.

As a government agency, MCHB is also formally required to provide accountability measures under the terms of the Government Performance and Results Act of 1993 (P.L. 103-62). These, too, rely heavily on vital statistics data as inputs. The performance measures for the Title V block grant program include such entries as decreasing the incidence of low-birth-weight births and increasing the percentage of pregnant women who receive prenatal care in the first trimester. Demonstrating progress toward these objectives is of great importance to the agency in the annual budget process.

It was noted in discussion at the workshop that some individual states and large counties are in the process of developing similar systems of specific accountability and performance measures. Krieger said that Massachusetts' health department had convened a series of regional hearings around the state, underscoring the importance of and interest in vital statistics indicators at multiple geographic levels. Massachusetts has established a commission on health inequities, with one part of the commission's work specifically related to performance measures and their use in guiding fund allocations to hospitals and other service providers. Van Dyck commented that California has worked to enforce standardized reporting by individual counties (providing data in the same format), cognizant of the federal-state partnership performance measures that it provides to MCHB and other agencies. Standardized reporting not only eases the process of producing the reports to federal grant supervisors, but also serves as a quality improvement and assessment tool for the state. Van Dyck said that developments such as these have been particularly interesting given worries among the states when MCHB began making state-level performance data public on its website about 10 years ago: states were concerned that people were going to use the data for unfair comparisons or to try to make individual states look bad.

MCHB provides direct grant support to states to enable information system improvement. At the time of the workshop, van Dyck said that this work was initially focused on improvements in selected states. MCHB has provided funds to Massachusetts and New Jersey to help them implement a web-based electronic birth record system (with infrastructure improvements that also facilitate conversion to the 2003 revised standard birth certificate). MCHB has also initiated a broader project with the Indiana State Department of Health to develop a fully integrated child health system: combining collection and processing of vital records, early hearing impairment detection and intervention, newborn dried blood spot screening, lead screening, participation in the WIC program (Special Supplemental Nutrition Program for Women, Infants, and Children), and birth defects. The ideal for the project is a system that is not subject to the same kind of duplicative work and clashing standards that may result when systems operate in an independent, stand-alone fashion. Van Dyck said that Indiana has also linked this system with the Indiana Health Information Exchange, with the additional benefit of providing a comprehensive electronic medical record on Indiana children to their medical providers.

## 2–B POPULATION PROJECTIONS AND ESTIMATES

Samuel Preston (University of Pennsylvania) opened the workshop session on the use of vital statistics in making population and fiscal projections by observing that projections are useful in understanding the future of the U.S. population—not as some remote abstraction but as estimates with very significant consequences. In particular, projections are important to assessing the future solvency of entitlement programs such as Social Security, which is required by law to be in close actuarial balance over a projected 75-year period. As another example, he noted the important social rami-

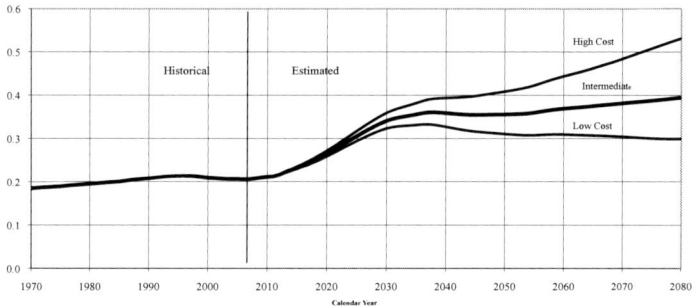

**Figure 2-1** Population aged dependency ratio, historical and projected through 2080

NOTE: Ratio calculated by dividing population ages 65 and older by population ages 20–64.
SOURCE: Workshop presentation by Goss.

fications of population projections, commenting that U.S. national identity is to some extent affected by projections that the majority of the national population will be nonwhite by 2050.

In addition to summarizing the projections program of the Social Security Administration, this section also consolidates material from two presentations by Census Bureau staff. The issues and challenges faced by the Census Bureau in producing current population estimates (between decennial censuses) are similar to those confronted in making future population projections. Hence, the two workshop presentations by Census Bureau staff on projections and estimates are most naturally summarized together here, even though they fell into different sessions at the workshop.

### 2–B.1 Population and Fiscal Projections at the Social Security Administration

The population projections made by the Office of the Chief Actuary (OCA) of the Social Security Administration are the driving force in determining the future revenues and costs of Social Security, Medicare, and other programs. Steve Goss (chief actuary, Social Security Administration) began his remarks by describing his office's use of the population aged dependency ratio as an explanatory measure; see Figure 2-1. The total dependency ratio, which would also reflect persons under age 20, is useful as well but—particularly for considering ramifications for the major entitlement programs—the aged dependency ratio is most useful for assessing the relative cost of programs.

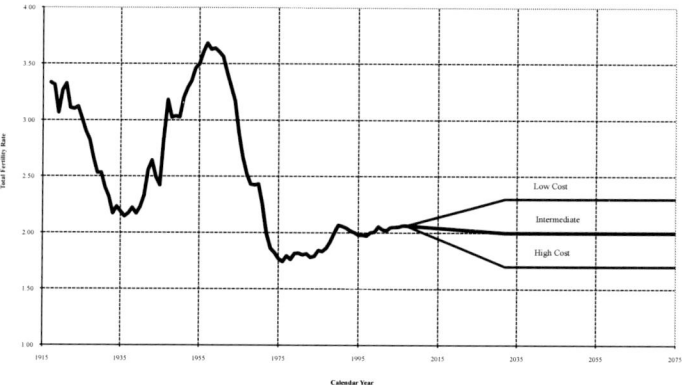

**Figure 2-2** Historical and projected total fertility rates, 1915–2075

SOURCE: Workshop presentation by Goss.

On the basis of historical and projected levels in Figure 2-1, Goss said that the arrival of the baby boom generation into retirement ages is a shift to a different level than the historical norm. The "intermediate" projections represent the office's best-guess assumptions, and those suggest a fairly level trend after 2030. However, the major level shift that the post-2030 estimate represents is a fundamental change in the age structure of the U.S. population. The shift is also a costly one, in terms of planning services to older populations. Goss further demonstrated that the basic shape and implications of the curve in Figure 2-1 is evident in other formulations: a modified version of the aged dependency ratio that also reflects some economic parameters, the ratio of the estimated number of beneficiaries per 100 workers, and the cost of Social Security as a percentage of taxable earnings.

Goss commented on OCA's use of and reliance on data on five different components of demographic change: (1) births and fertility, (2) immigration and emigration, (3) disability, (4) deaths, and (5) marriage and divorce.

## Births and Fertility

Goss commented that broad historical variability in fertility rates is well known; see Figure 2-2. Before about 1965—the end of the baby boom generation—estimates suggested an average of about three children per mother over the course of a lifetime. That average level has transitioned to a lower level, to an average of around two children per mother, since 1990. Goss said that many explanations have been offered for the drop. Analysis of data on U.S. birth rates by maternal age shows declining rates for women in their 20s but increasing rates for women in their 30s—part of a longer-term trend toward higher average maternal age at birth. Goss said that OCA understood that much of the 1965–1990 decline in the fertility rate was likely attributable to this transition to birth at higher ages and, consequently, not one that would continue to drop forever. Hence, Goss said that OCA has never assumed birth rates lower than 1.9 for the total fertility rate. Though some European countries do project a continued decline in fertility rates, OCA generally assumes a steady, average fertility rate of 2.0 for the U.S. population into the future.

Clearly, Goss said, the cost implications of shifts in birth rates for Social Security are substantial. The range of OCA's current projections at the end of 75 years—a total fertility rate estimated at 2.0, within an interval of 1.7 to 2.3—maps to a estimated cumulative effect of about 15.5–19.8 percent of payroll. That is, Social Security would require somewhere between 15.5–19.8 percent of total payroll earnings in order to pay all of its scheduled benefits. Goss demonstrated that changing fertility assumptions even slightly can have major effects on the estimates (and on the uncertainty relative to those estimates) of Social Security's funding needs.

Goss noted that OCA acquires its birth data from the NCHS-compiled vital statistics. In terms of data quality, Goss said that OCA is always concerned about the potential for underreporting, given the potential for distortion of the basic fertility rate that underlies so much of Social Security's fiscal projections.

**Immigration and Emigration**

Goss said that OCA resolves migration into four basic components and draws its data from a variety of sources.

- *Legal immigration:* OCA uses data from the U.S. Department of Homeland Security (DHS) on legally admitted immigrants by age and sex. OCA typically bases its assumptions on averages of these data over the past 10 fiscal years. Though most of the categories of legal immigrants are numerically limited or capped by law, one category that is not numerically limited is new immigrants who are immediate relatives of citizens. From its discussion with DHS staff, OCA has concluded that this category has been growing. Reconciling this information with some shifts in other categories (e.g., an increased tendency for persons acquiring legal permanent resident status to be people adjusting their immigration status rather than new entrants), OCA raised its standard assumption of 800,000 gross legal immigrants per year to 1,000,000.
- *Legal emigration:* OCA uses historical estimates of legal emigration produced by the Census Bureau, which have ranged from 20 to 30 percent of the level of legal immigration. OCA's current assumption

for this category is 25 percent of the level of immigration. However, OCA does make some adjustments to this working rule. In particular, people can leave the borders of the United States but retain their insured status for Social Security benefits; hence, for purposes of projecting Social Security needs, OCA needs to recognize this group in its calculations. Accordingly, OCA lowers its assumed number of emigrants at older ages—effectively treating them as non-emigrants for estimation purposes.

- *Other immigration (undocumented and temporary):* Historically, OCA relied on estimates of net immigration of U.S. residents. However, starting in 2008, OCA began working with separate estimation of both inflows and outflows in undocumented and temporary residents, with separate age structures. OCA's new calculations are based on analysis of 2000 census data, combined with estimates generated by DHS in 2006; the age distribution at entry (and exit) is based on unpublished Census Bureau tabulations for the net "other immigrant" count for 1975–1980. On the basis of this work, OCA's current annual assumption is about 1.5 million new other immigrants per year.

- *Other emigration:* On the basis of OCA's analysis, the office assumes that about 0.5 million of the 1.5 million other immigrants each year become legal permanent residents within 5 years. The other 1 million either stay (in undocumented or temporary status) or emigrate; OCA currently assumes that about 700,000 of that 1 million eventually exit the country.

In discussion at the workshop, Goss noted that, in making its projections of undocumented immigrants, OCA has to make assumptions about the extent to which the undocumented immigrants work for wages and, if they do, whether they pay taxes. OCA's current projections are that about half of new undocumented immigrants do pay into the system (Social Security and other taxes) but that the fraction will decline over time. In part, Goss said, this is due to the increased documentation requirements to obtain a Social Security card. OCA currently projects that only a relatively small fraction (10–20 percent) of undocumented individuals will go through the process of acquiring legal residence and actually receive benefits.

Goss commented that the implications of immigration for Social Security projection are relatively modest, with only about a 1 percentage point swing in the Social Security cost rate over the 75-year projection period being attributable to immigration.

### Disability

Though not commonly thought of as a vital event in the usual sense, Goss noted that disability is certainly an important and life-changing factor—with real implications for the cost of Social Security—and so is incorporated into the fiscal projections. In the absence of firm national data on disabilities, the Social Security Administration draws its data and assumptions on disability from its internal data. Specifically, OCA draws on Social Security data on incidence (based on entitlements and awards) and reported medical and work terminations.

### Deaths

For data on deaths, OCA augments NCHS-compiled vital statistics with Medicare data. For deaths of persons under age 65, the vital statistics of death by cause are the exclusive source, with Census Bureau population estimates as the denominator. For persons age 65 and over, Goss said that OCA tends to work with its own statistics, based on Medicare enrollments; although these data are limited to those people who are fully insured in the Social Security system, OCA has concluded that this approach gives it consistency in the numerator and denominator used in death rates and, moreover, helps minimize misstatement of age at time of death (as is a lingering concern with death certificate data). However, the vital statistics mortality data for persons age 65 and older are still an important input through their information on the distribution of death by cause.

Goss observed that OCA's death rates are projections by specific causes of death. To make such projections, Goss said that OCA pays careful attention to historical trends in mortality, but its final assumptions may reflect slightly differing expectations. Though mortality has historically declined rather rapidly at young ages and not very much at older ages, OCA tends to assume more rapid acceleration of mortality for the oldest ages (85 and above) than some figures would suggest.

Goss indicated that the cost sensitivity of its fiscal projections to assumptions on mortality is very significant—about a 3 percent swing over the 75-year projection period between its low- and high-end projections. However, fertility measures remain the most sensitive part of OCA's overall projections.

### Marriage and Divorce

Marriage and divorce are critical to consider because of their effects on both benefits and employments. Though NCHS no longer compiles marriage and divorce data in the national vital statistics, OCA continues to base it assumptions on age distributions from the last available national numbers—1995 for marriages and 1988 for divorce. OCA believes that it has a reasonably good handle on the total number of marriages and divorces, and its projections for both marriage and divorce are effectively flat, constant trends. Still, Goss said that OCA would clearly benefit from more recent and detailed data of the form that used to be compiled in the national vital statistics program.

### 2–B.2 Population Estimates and Projections at the Census Bureau

Victoria Velkoff and Fred Hollmann (both of the U.S. Census Bureau) addressed the workshop on the use of vital statistics data in the Census Bureau's intercensal population estimates and its projections of U.S. population trends.

By law (13 USC §181), the Census Bureau is required to produce basic estimates of population and demographic characteristics between decennial censuses:

> During the intervals between each census of population . . . the Secretary, to the extent feasible, should annually produce and publish for each State, county, and local unit of general purpose government which has a population of fifty thousand or more, current data on total population and population characteristics. . . .

Velkoff commented that the population estimates are used to allocate over $300 billion in federal funds each year, and they are also used by some states in their funding formulas. The Census Bureau also uses the intercensal population estimates as controls or weights in major household surveys such as the Current Population Survey, ACS, the Survey of Income and Program Participation, and the American Housing Survey. The Bureau of Economic Analysis uses the population estimates in its estimates of per capita income, and they play significantly into calculations by other federal agencies. At the workshop, Kenneth Prewitt (Columbia University) pointed out one particular federal use that illustrates the circularity and feedback loops in the broader statistical system: because vital statistics on births and deaths are a critical component of the population estimates, vital statistics drive both the numerator (incident counts) and denominator (population) of NCHS's calculated birth and death rates.

As the Census Bureau's system has evolved, national- and state-level population estimates are released by the end of the reference (estimate) year. Estimates disaggregated by demographic groups and for smaller geographic areas are rolled out over the course of the year; Velkoff noted in particular that the Bureau's nation and state demographic estimates for specific demographic categories for 2007 were slated for release the day after the April 30, 2008, workshop.

The Census Bureau's population estimates program generates on an annual basis:

- national-level estimates, total and disaggregated by age, sex, race, and Hispanic origin categories;
- state-level estimates, total and disaggregated by age, sex, race, and Hispanic origin categories;
- estimates for the 3,141 counties (or county-level equivalents), total and disaggregated by age, sex, race, and Hispanic origin;
- estimates for about 39,000 incorporated places (cities and towns) and minor civil divisions (county subdivisions), total population only; and
- estimates for Puerto Rico and its county-level municipios, by age and sex.

Consistent with the U.S. Office of Management and Budget (OMB) guidelines, the full disaggregation by race and Hispanic origin involves 62 categories: the 31 combinations of the five race categories crossed with two Hispanic origin categories (Hispanic or not Hispanic).

Velkoff described the basic cohort component method used to generate Census Bureau estimates: updating the most recent decennial census count by adding births, subtracting deaths, and adding an estimate of net international migration. The Census Bureau generally relies on matches of Internal Revenue Service records from year to year to estimate domestic migration (supplemented by Social Security and Medicare data) and the Bureau's own ACS for estimating international migration rates. The NCHS vital statistics data are the basis for the estimates of births and deaths used in the cohort method.

The Census Bureau also periodically releases long-term population projections to describe the demographic character of the future U.S. population. Hollmann said that these projections are used by states and localities for specific planning objectives, such as assessing demand for roads, schools, and other infrastructure improvements. Among federal users, the Bureau of Labor Statistics uses the population projections as the basis for its own projections of the future labor force, and the National Center for Education Statistics uses them to plan education estimates.

The population projections program does not approach the level of geographic detail of the population estimates program; it produces only national- and state-level projections (though Hollmann noted that projections for metropolitan areas are sometimes discussed as a future improvement). Like the full suite of population estimates, the release of population projections is also staggered over time (albeit a longer time range than the annual estimates): At the time of the April 2008 workshop, Hollmann indicated that the most recent (interim) national projections were released in

2004, the most recent projections of total state populations dated to 2005, and national estimates disaggregated fully by race and Hispanic origin categories was slated for release in summer 2008.

Until recently, the Census Bureau's population projections were particularly dependent on input from vital statistics because they relied on an essentially census-independent population base. The projections program relied on the so-called demographic analysis population based on historical vital statistics and estimates of international migration. However, Hollmann said that the Census Bureau abandoned that practice in 2000 as a result of comfort with the smaller level of aggregate undercount in the census. Hence, the Census Bureau is using a census base for its rates and projections rather than a purely demographic basis.

Still, the Census Bureau's population estimates and projections both hinge on the NCHS-compiled vital statistics to provide birth and death components in various formulas. Velkoff and Hollmann both commented on the experience of using the vital statistics data and analyses:

- Velkoff noted that the time lag in the availability of final vital statistics raises some concerns for the Census Bureau's work. Given the Bureau's internal timeline of producing some national estimates by the end of the target year, the Census Bureau typically finds itself in the position of "projecting" vital events for about a year and a half. Although Velkoff indicated that the Census Bureau has adjusted to this situation and that these internal projections typically turn out to be of acceptable quality, there are some instances in which they do not—the significant population shifts due to Hurricane Katrina in 2005 being a prominent example. For the purposes of computing the estimates, Velkoff said that the Census Bureau makes the assumption that local reporting of births and deaths is 100 percent complete; however, Census Bureau researchers who develop the Bureau's demographic analysis estimates have conducted studies that relax the assumption of constant, complete reporting. The Census Bureau has also recently become concerned with the quality of data on reported age at death, particularly at the oldest ages, and is conducting work to evaluate its internal model for mortality at the oldest ages.

- Citing Census Bureau research on the components of change within the national population projections (Mulder, 2002), Hollmann commented that the Bureau has found that, historically, the largest source of error in its projections comes from projecting fertility. He noted that this is not to say that the Bureau has made the largest-magnitude errors in projecting fertility rate, but rather that variability in the birth component tends to have the largest effect on the final estimates. Over the long term, Census Bureau projections both overprojected fertility (e.g., in the late 1960s and 1970s on the back side of the baby boom) and conservatively underprojected fertility (in later years). Errors have tended to be less acute in projecting death and international migration (even though the Census Bureau generally projects the latter as a constant, and thus is almost certainly consistently low).

Both Velkoff and Hollmann commented on the challenge of using vital statistics data in their current methodologies, given current variation among registration areas in the format of race and Hispanic origin data; this specific methodological discussion is summarized in Section 4–B.

Discussion at the workshop centered on the trends projected for the Hispanic population and, in particular, on research on differential trends in mortality and general health among Hispanics by generation. Nancy Krieger (Harvard School of Public Health) asked whether nativity in the United States, versus foreign-born, is factored into the projections. Hollmann indicated that neither nativity nor specific type of Hispanic origin (e.g., Mexican or Cuban) is directly factored into the models but that it is picked up to the extent that the projections are based on a historical series driven in part by such differential trends. Session moderator Samuel Preston (University of Pennsylvania) commented that lower mortality rates among Hispanics are evident in Social Security Administration data, which may be less immune to data reporting effects in the vital statistics data (i.e., without birth and death certificates playing such a major role in both the numerator and denominator of calculated rates).

Since the early 1970s, the Census Bureau has worked with a network of state agency contacts to assist in the production of the annual estimates; later in the decade, this Federal-State Cooperative Program for Population Estimates (FSCPE) was joined by a companion Federal-State Cooperative Program for Population Projections (FSCPP). Under both programs, the states designate an agency as their representative. In some cases, Velkoff noted, the state FSCPE contacts directly provide vital statistics to the Census Bureau (particularly inasmuch as some states' designated FSCPE or FSCPP agency is also its participant in the Vital Statistics Cooperative Program).

## 2–B.3 Discussion

In discussion of this session of the workshop, moderator Samuel Preston (University of Pennsylvania) commented that one of the values of vital statistics, as they are produced for analysis of fertility and mortality, is that they can be arrayed by birth cohort. Cohort analysis suggests interesting patterns in both fertility and mortality that are not possible to observe by just considering period behavior. For births and fertility, no individual cohort was as extreme in magnitude as the peak in the (period-based) total fertility rate would suggest, and the cohort-level trends have less variability than in period behavior. For deaths and mortality, cohort mortality patterns raise interesting findings looking as far back as the 1930s and the first studies of cohort mortality patterns. In particular, when sex differentials in mortality are arrayed on a cohort rather than a period basis, the sex differential peaks for the cohort born around 1905—exactly the cohort for which sex differences in smoking behaviors also reached a peak.

Goss concurred that this suggests that vital statistics—and cohort analyses of them—provide great opportunities for investigation of patterns in mortality and fertility. Goss said that significant work had been done internationally on this, particularly in the United Kingdom. Goss said that OCA had also compared its data with counterparts in Canada; Canada has not seen the same broad improvement in its national rate of mortality as in the United States, but further analysis of trends could be useful.

## 2–C  GROWING AND EMERGING USES: VITAL STATISTICS AND BIOSURVEILLANCE

A workshop session moderated by Kenneth Prewitt (Columbia University) considered important applications of vital statistics beyond the health care planning domain. Michael Stoto (Georgetown University) spoke of his recent work in health surveillance for national security, also known as syndromic surveillance or biosurveillance. Although originally focused on the detection of terrorist attacks using biological agents, Stoto argued that biosurveillance has come to be interpreted more broadly, as a means for situational awareness for public health emergencies. In either event, Stoto noted that the data systems he was discussing have a much more exacting standard for timeliness than the current vital statistics collections—timeliness measured in weeks and days, and sometimes hours, rather than years. Indeed, the basic point of near-real-time acquisition and use of prediagnostic health data is that waiting until people are diagnosed with diseases or, in the case of vital statistics, die from them would be too late to inform possible interventions. Still, he said, there are important linkages between biosurveillance and the current vital statistics.

The central statistical challenges in biosurveillance are, first, obtaining and integrating accurate data from a variety of sources in a timely way and, second, determining whether something is "unusual." The latter task is complicated by high variability in the background and a possible unstable process generating the data; it involves making critical trade-offs among sensitivity (i.e., false negatives), specificity (i.e., false positives), and timeliness.

Current work in biosurveillance has sought to build on existing data systems in the health care world—such as emergency department reports, sales of over-the-counter medication, and absenteeism from work and school.

These data are usually electronically gathered and highly networked. Using these data, statistical analyses can detect sudden changes that might suggests a disease outbreak or maybe a covert bioterrorist attack. As an example, Stoto described analysis of emergency room data for seven Washington, DC–area hospitals during winter 2003; for those data, detection algorithms suggested certain excesses of gastrointestinal diseases at several hospitals at several points (e.g., early February, early March, mid-April). Though not definitive—Stoto said that there is no way to know whether, for example, the early February increase is a sudden escalation due to random chance or the beginning of something big, that all that is known is that there are differences from what happened before February 1—the work provides clues to follow in ferreting out causes.

Biosurveillance research also involves dealing with a number of practical issues. These include privacy concerns about the patients to whom the data refer, as embodied in the Health Insurance Portability and Accountability Act Privacy Rule and related state laws. Other practical challenges include proprietary concerns (who owns the data and with whom can they be shared), concerns and possible prohibitions on secondary use of the data, and operational costs for personnel and information technology. These formidable practical challenges have partially contributed to the recent shift to cast biosurveillance more broadly as a "situational awareness" technique, Stoto said. Though the purely statistical questions of detecting significant increases in activity continue to draw attention, interest has shifted toward public health activities such as "case finding": making it easier for physicians and other health care providers to report individual cases that might be of concern and when and where something might be going on, as well as ways to aid outbreak investigations and monitor outbreaks.

Given the emphasis on timeliness that is central to biosurveillance systems, Stoto asked what kind of contributions vital statistics—and the principles and practices of vital statistics systems—can bring to bear. Mortality due to pneumonia and influenza is an instructive example to consider for a number of reasons, among them a lengthy history of analysis of such data, experience with the challenge of distinguishing between routine seasonal influenza and wider pandemic outbreaks, and the fact that exposure to many biological agents that could be used in a terrorist attack would initially cause flu-like symptoms. Stoto began by noting recent work by Mills et al. (2004) analyzing data concerning the 1918 Spanish influenza epidemic in the United States. That research found big differences from city to city in terms of the timing and extent of excess influenza deaths. Stoto said that having such data available during an epidemic would certainly have been useful. Close analysis of data (Collins et al., 1930) suggests further insights on public response to the disease outbreak. Stoto noted that a classic example that has been raised from the data is the difference in response by two cities—Philadelphia, which did relatively little in response, and St. Louis, which was much more interventionist in terms of closing schools and reducing public activities.

Pneumonia and influenza mortality data for 122 cities that have relatively rapid vital statistics reporting are still a key component of public health influenza surveillance. Like the historical data, these data provide useful insight on the timing and extent of flu in one year relative to previous years. Timeliness, however, is a problem: Data downloaded on one Saturday refer to deaths in the week ending the previous Saturday. Yet death from influenza is the end point of a process that typically runs from 1 to 2 weeks, and so even with the most timely data, insight on the time and nature of infection is limited by an effective 3-week lag.

The example of influenza monitoring raises the question of how modern information technology—such as electronic vital records collection and electronic death records—might make mortality data more useful for near-real-time monitoring. Vital statistics may never achieve the near-hourly temporal resolution that is needed for outbreak detection, but Stoto suggested that there is still a great deal of value in being able to frame assessments based on what is going on in cause-specific mortality data on a monthly or weekly basis. The question of geographic representation is an important issue: the flu surveillance data from which Stoto drew in his example is based on the cities that, more than a decade ago, were at the forefront in being able to gather electronic death records; more development, and assessment of the coverage represented by such cities and areas, is essential.

Stoto closed by noting that vital events reporting had advanced technologically—going beyond the postcards used for reporting in the 19th century to data compilation by fax, phone, and the Internet. However, in its basic character, the collection of information on vital events and disease has not progressed much from the old postcard system. He argued that there is great benefit in the basic structure of the current vital statistics system, which leaves "ownership" of case reports with the state and local authorities, but which provides for federal ownership of the system for gathering vital records and compiling them. Stoto urged that the nation consider a notifiable diseases cooperative program, akin to but broader in scope than the existing vital statistics cooperative program.

Ed Hunter's presentation (Centers for Disease Control and Prevention) focused principally on the demands on birth certificates and the issuance process due to antiterrorism legislation (see Section 3–A), but he concurred with key parts of Stoto's presentation. Hunter said that it is still unclear whether birth certificate requirements will be the final major impetus for a fully electronic, rapid, standardized vital registration system for births and deaths (given legally mandated matches of birth and death records). If it is, however, Hunter said that the system improvements required by the security provisions will do all the things that Stoto mentioned were necessary for situational awareness and surveillance, including rapid availability of vital records data and, ideally, marked decreases in the time lags to issuance of birth and death data. This development would also have a variety of beneficial spillover effects for the general study of health information. Hunter concurred in the usefulness of pandemic influenza as an initial study and development area; he said that the pandemic funding stream is another opportunity to build on the state electronic registration systems and to try to advance their timeliness.

# – 3 –

# The Federal-State Cooperative Relationship

WITH ITS MANDATE TO COLLECT DATA on vital events, the National Center for Health Statistics (NCHS) shares a common challenge with several of its peers in the highly decentralized federal statistical system: functioning as a national-level collector of information on phenomena that are inherently local in nature. Accordingly, mechanisms for cooperation and coordination between federal statistical agencies and state or local authorities are common in the statistical system, ranging from relatively simple awareness-building activities (e.g., the multitude of short-term partnerships that the Census Bureau forges to boost participation in the decennial census) to highly structured contractual and financial agreements (e.g., the grants administered by the Bureau of Justice Statistics to support development of criminal history record databases).

The Vital Statistics Cooperative Program (VSCP) that has been formed between state and local registration areas and NCHS is, as workshop presenter Ed Hunter (Centers for Disease Control and Prevention) noted, a federated system, with the national or federal-level entity of NCHS providing funding, coordination, and standards, but with the individual states and localities retaining significant autonomy in their operations. The general structure of the system and the special challenges it faces—some challenges common to nearly all federal data collection efforts in a time of scarce resources, but others unique to the nature of the vital records that are the source of data in the VSCP—was a recurring theme at the workshop.

In this chapter, we summarize some threads of this discussion on the structure of the VSCP. Section 3–A describes the role of the states and, more generally, the challenge of data collection given the civil registration nature of the underlying birth and death certificates. Section 3–B summarizes the constraints on the vital statistics collection system from NCHS's perspective as national-level coordinator. Finally, Section 3–C profiles selected models of federal-state cooperation elsewhere in the federal statistical system.

Workshop presenters and participants received two background papers prepared at the workshop planning committee's request: one on the role of the states (prepared by Steven Schwartz, New York City Department of Health and Mental Hygiene) and the second on NCHS's role. These papers are reprinted in Appendixes A and B, respectively; they have only been minimally edited for consistency with National Research Council report style. Schwartz gave a presentation at the workshop that closely followed the content of his paper; there was no explicit counterpart presentation of the background paper by NCHS, but several of the paper's themes were sounded by Jennifer Madans and other NCHS staff (as summarized in Section 3–B).

### 3–A  THE ROLE OF THE STATES

Schwartz emphasized the local nature of vital events and the collection of records. At the outset, records of each of the more than 11 million vital events—births, deaths, marriages, and divorces—that occur each year in the United States are processed through one of the more than 6,000 local registrars. These local registrars form a diverse and complex network, and Schwartz's own local experience offered an interesting perspective on the geographic distribution and the workload of registrars. The state of New York has the most local registrars of any state—about 1,500—yet New York City and its more than 8 million inhabitants have only one official registrar. The New York City registrar's office alone processes about 500 live births and 160 deaths every day—about 300,000 vital events annually.

Records data funnel through the local registrars to the 57, mainly state-level, registration jurisdictions. As already noted, two cities—New York City and Washington, DC—function as registration jurisdictions. (Schwartz observed that, at one time, registration districts were more city based and, in fact, New York City began as a registration area before New York state.) Each of the 57 registration jurisdictions reports data directly to NCHS through the VSCP. Schwartz also noted another centralizing force in the system—the National Association for Public Health Statistics and Information Systems, the professional association of the state vital records offices that was founded in 1933 and works with NCHS and the states on data collection issues.

Schwartz observed that the registration jurisdictions vary considerably in their capacities and individual procedures, their staff size and expertise, and their level of system automation and electronic database implementation. He said it is important to bear in mind that the state and local vital records offices must fulfill three basic—and sometimes competing—roles. Two of these roles have historically dominated the work of the offices:

1. *Civil registration of vital events:* The most basic function of the vital records offices, the civil registration function of vital events, has the important implication of making the offices huge customer service operations. Schwartz said that walk-in customers have pressing legal needs for records and certified copies and may require corrections or amendments to existing legal documents; as custodians of the records, the vital records offices also have a responsibility to be prompt and responsive. Schwartz noted that his New York City registrar's office fields on the order of 700 walk-in customers a day, seeking copies of records or other services; the approximately 800,000 paid copies of birth and death certificates that the office issues each year accounts for about $12 million a year in revenue.

2. *Public health statistics collection:* It is the processing of records data—the information on the birth and death certificates—that ultimately populates the vital statistics data files. Schwartz noted that there is a constant tension in resource allocation between the statistical role and the civil registration and customer service role and that the statistical side must take second place.

Yet a third and no less important role has arisen and been made explicit in law in recent years: vital records offices are also front lines in *national security* efforts. Both Schwartz and Hunter noted that birth certificates have become particularly sensitive because they are breeder documents that are the basis for many other important documents and legal statuses. Birth certificates can constitute proof of U.S. citizenship; because each vital records reporting jurisdiction maintains a contract with the Social Security Administration (SSA), birth certificates also trigger issuance of Social Security numbers and eligibility for benefits. Birth certificates are also used to obtain state driver's licenses and federal passports, key means of establishing identity.

The Intelligence Reform and Terrorism Protection Act, which became law in December 2004, was a partial implementation of the recommendations of the National Commission on Terrorist Attacks Upon the United States (2004). Among its provisions was a set of minimum standards meant to secure birth certificates. Hunter argued that the changes in the 2004 act were not really new—a 1996 immigration act passed by Congress contained many similar provisions. However, the 2004 act was substantively different in important ways and carried particular urgency given that several of the terrorists in the attacks of September 11, 2001, had obtained passports and identity documents using fraudulently obtained birth certificates. Recognizing the federated nature of the vital registration system—not national or federal per se, and so lacking the ability to directly effect changes—the 2004 bill simply stipulated what the federal government would accept as a valid birth certificate. This left implicit—and up to the states and localities—what changes to the system were needed to meet those standards in order for certificates to be valid for federal purposes. As Hunter summarized, and Schwartz echoed, the demands of this new national security role for vital records offices are considerable:

- The 2004 act sought to reduce the hundreds of different variations of birth certificates, including commemorative or ceremonial ones issued by the states, that were previously allowed. The act defined standard, recognizable paper certificates, printed on a certain type of security paper, as well as other provisions for basic structure.

- In addition to securing the physical document, the 2004 act's provisions are intended to secure the system by which they are issued. One part of this revised system is a requirement for rapid ascertainment of death certification and a direct matching of death to birth records to preclude fraudulent use of birth certificates of the deceased. Like other background check systems, the critical requirement of this matching is that it needs to cross all jurisdictions and it needs to be fast. However, although the act was signed in December 2004, Hunter's workshop presentation described "regulations for secure systems" as "pending" and still in progress; NCHS put draft regulations together, but these are still circulating through the federal system.

- Although the act's text only explicitly speaks to standards for birth certificates, the requirement of matching of birth and death records tacitly also implies standards for death certificates.

Schwartz noted that the new national security role is one that is resource intensive for localities and, again, one that can blur the states' ability to focus on the public health data collection role. In discussion, Kenneth Prewitt (Columbia University) noted another way in which the security role potentially clashes with the data collection role. He worried that, to the extent that electronic vital registration systems become portrayed as a homeland security tool or even a type of law enforcement mechanism for detecting fraud, complications may arise for statistics. That is, security may become so tight and participation sufficiently strained that it may be more difficult to move the system into the kind of social, public health surveillance system that is needed for detection of early disease or other health incidents.

In terms of the functioning of the VSCP, Schwartz expressed strong support for the basic distributed nature of the system. The locally distributed system of collecting vital event data gives state and local registrars maximum leverage to work closely and directly with their source data providers: hospitals and physicians, funeral homes, nursing homes and clinics, and so forth who contribute the information that populates vital records. This approach satisfies local authorities' need to effectively be the master of their own data stream—to best know their own data and their own data providers.

However, he also agreed that it is not a perfect system, and noted several particular challenges:

- At the national level, the compiled vital statistics data can only be as timely (and as high quality) as the weakest state. Reporting lags and data quality or consistency issues of an individual state, large or small, can impair the national system.

- A related challenge is that the distributed nature of the VSCP makes training, educating, and querying of the source data providers a key aspect of improving and maintaining end data quality, particularly the case for cause-of-death reporting, for which consistency in approach is critically important. However, such training and education efforts are probably the first thing to suffer in a competition for scarce resources.

- The individual registration jurisdictions are sensitive to the amount of funds that NCHS provides through the VSCP, and uncertainty over NCHS's resources can have significant local effects. In discussion, Schwartz commented that New York City's use of NCHS-provided VSCP funds is principally to pay staff. He noted the example of NCHS's discontinuation of abortion reporting in 1995, the result of which was that New York City lost more than $75,000 in funding and had to forgo a staff position and active surveillance of abortion providers. Hence, in recent reports, the city has had to attribute drops in the number of abortions to the cessation of active monitoring—rather than a real decline—because its ability to accurately measure activity has been impaired.

- A further complication in the distributed nature of the system was suggested by then–Census Bureau director Steve Murdock in his luncheon remarks at the workshop, drawing on his experience as state demographer of Texas. In states with extensive rural populations, such as Texas, the county clerks responsible for processing birth and death certificates (not to mention other government documents) may be part-time positions and, hence, data collection and processing (and furtherance of the national vital statistics collection efforts) might not be a high priority. He said that the part-time nature of these jobs contributes to the growing pains that occur as paper-and-pencil registration systems become computerized systems and to the lags that result in some areas.

Like Hunter, Schwartz commended the development of electronic systems for the automated processing and verification of birth and death records. The system has made great strides in moving away from having the data on paper forms key-entered by local offices; web-enabled systems that permit hospitals and doctors to enter and certify data electronically have the potential for improving data quality and timeliness. The catch, he noted, is that these automated systems carry extremely high start-up costs—on the order of $1 million per death or registration system—that state and local registration offices have had difficulty obtaining from their parent governments. The result—as noted by other workshop presenters—is a lack of uniform implementation. Schwartz's paper in Appendix A provides additional details on specific electronic systems that have been developed or proposed, including the national State and Territorial Exchange of Vital Events that is proposed to permit matching of birth and death records across all vital statistics offices.

Going forward, Schwartz suggested that it is important for the stakeholders in the VSCP to consider ways to look at the system's return on investment. He argued that the system is not broken but that there is no clear measure of how good or how bad it is: some estimates of the return on investment of resources at the national and state levels would be valuable for building support for the system among policy makers. He suggested the example of the SSA's Enumeration at Birth Program (EAB; initiated in 1990) that assigns Social Security numbers to newborns, with parental approval, when a birth certificate record is processed. The SSA Office of Inspector General audited the program (using SSA's own data) and estimated the average cost of the traditional process—individual parents walking into local Social Security offices to obtain a number for their newborn children—to be $18.70 per record process; in comparison, the audit suggested that the total cost of processing records under the EAB system (including the fee that goes to the states for processing) is only $3.74. As Schwartz noted, the resulting estimate of about $60 million in savings each year (about $15 per record, multiplied by about 4 million births per year) is a clear and compelling assessment of the return on investment of the EAB program. Noting that the EAB program might be a relatively easy-to-measure case as a pure administrative system, Schwartz argued that these kinds of figures are worth considering in relation to vital statistics. Measuring the value of vital statistics is considerably harder—trying to quantify things such as the value to research of the data files and the returns of the use of locally held vital records data to populate immunization or lead-exposure registries or to target newborn home visits by nurses. Still, such measures are important to consider in making the overall VSCP better and demonstrating its unique value.

## 3–B CHALLENGES AND LIMITATIONS AT THE NATIONAL CENTER FOR HEALTH STATISTICS

The largest problem facing the current VSCP from NCHS's perspective was first raised very early in the workshop. Committee on National Statistics director Constance Citro observed that the workshop was planned with three Cs as themes: to *celebrate* the many and growing needs served by vital statistics, to *critique* the program to identify strengths and weaknesses, and to *contribute* to a better understanding by policy makers of the value of vital statistics. NCHS director Ed Sondik countered that the third of these Cs could readily be simplified to *"costs"*—grappling with the continuing challenge of obtaining high-quality data through a cooperative program when both federal and state resources are tight. He commented that in order for the center to balance the agency's budget, NCHS does not have sufficient resources to fill all of the vital statistics needs. Indeed, he commented that NCHS has been meeting with its Board of Scientific Counselors specifically to discuss options for the programs should NCHS's generally flat funding continue.

The presentation by Jennifer Madans (NCHS) in the final session of the workshop echoed the concern about costs. She observed that the desire to build the vital statistics system and revamp it—including further promotion of electronic systems—is coming at a time when both NCHS and state governments are facing tight funding constraints. In its planning, NCHS has had to generally assume a flat budget going forward and simultaneously wrestle with the data collection costs for information collections in all other areas in health. Accordingly, she noted a certain level of frustration by all vital statistics stakeholders: there is a strong need for the data and great pride in the vital statistics system, but not a great deal of latitude for massive improvement in one single program without resources. The sensation is one of striving to meet today's problems by putting off tomorrow's problems, which are investments in future capacities: the problem is that sooner or later the VSCP is going to get to tomorrow.

From NCHS's perspective, a constant challenge given the cooperative nature of the VSCP is determining fair shares of costs. As mentioned above, the states and localities must deal with the civil registration aspect of vital events, and the staff and resource allocation to keep up with customer service is a very significant local administrative focus. The key questions are: What is the cost of gleaning the data from the records and the value of the information that the states and localities possess (both in local totals and compiled national data)? How do those values lead to a determination of who is responsible for what part of the overall costs?

In this context, Madans noted that the VSCP is an interesting case in point in the broader federal statistics system. In many agencies and applications, she said that a common theme is the use of administrative records and administrative data for several purposes. The argument is that the use of administrative data will make things different (ideally, better) and will require some changes in interpretation but will almost certainly make things more efficient and less expensive (to the extent that the administrative data are used to reduce field data collection). Vital statistics are commonly pointed to as a long-standing example of such administrative data being used. What is not always appreciated, though, is the synergy that went into the development of the VSCP. In other statistical applications, administrative data tend to be thought of as records that have been generated and created for a totally distinct purpose that can then be tapped at little or no cost and added as part of one's database. Because of the cooperative nature of the VSCP and the revisions of standard certificates, the system itself develops the form and content of the administrative record. This gives the system great flexibility in the content of the data—it is not clear what would be on birth and death records if this cooperative system had not developed—but carries with it significant costs and challenges.

## 3–C EXAMPLES OF FEDERAL-STATE COOPERATION IN THE U.S. FEDERAL STATISTICAL SYSTEM

Two workshop presenters described the general structure of federal-state cooperation within their agencies, to suggest possible improvements in the structure of vital statistics collection. The two specific systems considered in these presentations are of interest because of their parallels to the collection of vital statistics. The Quarterly Census of Employment and Wages (QCEW) parallels vital statistics in that the source data are essentially administrative records with other legal purposes (in this case, tax filings to unemployment insurance programs); it differs from the vital statistics program in having a strongly defined set of legislative requirements for the structure of the federal-state partnership enacted in recent years. The second example, the Education Department's Common Core of Data (CCD), parallels vital statistics in that initial responsibility for data completion is diffused among a wide variety of state and local authorities (in this case, individual elementary and secondary schools as well as state departments of education); it differs from the vital statistics in that several of the components of the CCD serve a primarily directory-building role rather than an analytical role.

### 3–C.1  The Quarterly Census of Employment and Wages

Jack Galvin described the cooperative structure underlying the QCEW program of the Bureau of Labor Statistics (BLS), which produces quarterly data on employment and wages at the national and subnational (state, metropolitan statistical area, and county) levels.[1] BLS's federal-state cooperative efforts date back to at least 1916 and were particularly strengthened by a provision of the Wagner-Peyser Act of 1933 that directed the Labor Department to reimburse states for the operation of statistical systems that contribute to national statistical series.

One of five formal federal-state cooperative programs currently maintained by BLS, the QCEW federal-state partnership is a relatively recent development at BLS, which assumed technical responsibility for the program from the Department of Labor's Employment and Training Administration in 1972 and financial responsibility in 1984. QCEW data are known for their comprehensiveness and for their ability to describe local-area economic conditions. Within BLS, QCEW data are also essential as a building block for other statistical programs (e.g., as a benchmark for the annual payroll survey and a sampling frame for establishment surveys), and they are also an important input to the national accounts studies of the Bureau of Economic Analysis.

Through the QCEW program, BLS directly funds its state partners to collect and edit data from the state-based employment insurance programs. As with vital statistics, the underlying data of the QCEW are, essentially, administrative records: the quarterly contribution reports that employers supply to the state-based employment insurance programs when they pay their taxes each quarter. The approximately 7.7 million quarterly forms, from over 9 million separate business establishments, are estimated to include coverage of about 98 percent of jobs in the United States. In compiling the data, the states are responsible for providing some information that is not directly coded on the quarterly tax forms, such as the industrial classification of the business and verification of physical location address (rather than general mailing address). To provide these supplemental data, QCEW funding requires the states to contact each business establishment every 3 years.

In addition to funding, BLS's role is also to provide technical and methodological direction to the data collection. It provides the information technology systems for the collection and processing of the data and promulgates standards for the data that are sent to BLS. Significantly, BLS plays no role in the structure or format of the quarterly contribution forms, and so the individual state employment insurance programs can vary in the form and filing requirements placed on employers (e.g., whether reports can be filed electronically or on paper). Responsibility for dealing with the variety of inputs from employers remains with the states; what BLS's QCEW funding promotes is a set format for the specific economic elements that are coded in the data files that are returned to BLS.

---

[1] Additional detail on the QCEW program can be found at http://www.bls.gov/cew/ (April 2009).

Galvin indicated that BLS spending on the QCEW program in 2008 was $49.5 million; the share of that total that is allocated to the states ($30.6 million) makes it the largest of BLS's federal-state cooperative programs. From 2000 to 2008, funding for the QCEW grew at an average of 2.9 percent per year, in line with the approximate 2–3 percent increase in the number of businesses in existence each year (and a corresponding increase in workload for the state officials). Galvin noted that BLS has typically been successful in securing such "mandatory" increases from the Office of Management and Budget and from congressional appropriators, giving QCEW relatively stable funding.[2]

As NCHS does with the state vital records offices, BLS negotiates individual QCEW contracts—cooperative agreements—with each state. The QCEW agreements are updated and agreed to on an annual basis. Through these annual agreements, BLS is able to specify (and modify, as appropriate) the expected quality standards for the data, the requirements for protection of confidentiality, and policies for allowable costs. By agreement with the states, BLS's funding to the states is calculated by multiplying a state's average government employee annual wage—itself a figure derived from published QCEW data—by 1.5 by the number of staff positions to be filled in the state. Each state is allocated a base of two positions for supervision (and continuity of operations), and additional positions are calculated on the basis of the state's workload in the program: for instance, its number of single and multiunit businesses and new business units.

Galvin indicated that BLS has concluded that its structuring of the QCEW program—and the stability of funding for it—has improved the quality of the data and improved the consistency of data quality across the states. The annual cooperative agreements provide a means to promulgate methodological standards, including routines for addressing situations found through review of edit failures in data processing and for the handling of missing quarterly tax reports (through imputation). BLS's provision of information technology to the partners also provides consistency and reliability in processing: Each state uses one of two technical systems (those used and developed in Utah or Maine), a considerable simplification from 50 heterogeneous technical systems. Galvin noted that the common technical platforms produced a significant benefit in terms of timeliness of data release: over the course of a couple of years, BLS was able to shift its quar-

---

[2]As noted in the discussion of Galvin's presentation, the use of "mandatory" means increases in funding because of increases in the cost of collection (e.g., cost-of-living adjustments to staff salaries), not that the increases are required by law. Galvin noted that BLS does not always receive these "mandatory" increases; in 2007, failure to get $2 million to cover the increased costs led BLS to cover the funding by other means, eliminating production of estimates from the payroll survey for several metropolitan areas and cutting sample size from another federal-state cooperative program that collects occupational employment statistics.

terly publication of QCEW data earlier by 3 weeks. By doing so, BLS was responsive to a major client for QCEW data—the Bureau of Economic Analysis, which (with data available that much earlier) can now use the quarterly data to update its personal income estimates four times per year rather than waiting to perform an annual revision.

Galvin noted that federal-state cooperation in the QCEW program does occasionally encounter vulnerabilities due to variation in state laws and regulations. State-level changes to filing practices for employers' quarterly contribution reports—the source data for QCEW—can create complications. In one instance Galvin cited, a state requirement that all employers begin filing the forms electronically (rather than on paper) proved difficult for small employers, contributing to glitches in processing and, for the QCEW, increased levels of nonresponse for that state. State confidentiality laws can also complicate the use of QCEW information for other purposes: For instance, they may prohibit BLS from sharing the administrative data collected in QCEW to assist the Census Bureau in filling in missing industry codes in its business and employment data.

In recent years, the QCEW Program has been further structured in response to the Workplace Investment Act of 1998. Section 309 of the act (P.L. 105-220) amended the original Wagner-Peyser Act to create specific provisions related to employment statistics. State governors were directed to designate a single state agency as the manager and coordinator of employment statistics for each state, but the act also established more direct oversight responsibilities for these state partners. The Secretary of Labor (and BLS) was directed by the act to develop a process by which 10 directors of these state-designated regions—one for each region—are elected to hold formal consultations with BLS on the cooperative management of the system.[3] This was a departure from established procedure in BLS's federal-state cooperative efforts such as QCEW, which had previously been handled more through BLS's regional offices than its national headquarters staff. In addition, the 1998 act directed that BLS hold formal consultations with the states at least once each quarter on the products and function of the employment statistics system.[4] Galvin noted that the organizational changes needed to comply with the 1998 act were difficult for BLS, given the entrenched practices, but that, by improving communication and feedback, they have improved the program and made it easier to accomplish further change in specific federal-state efforts.

---

[3] BLS has settled on a policy by which these elections are held every 2 years.
[4] Specifically, the mechanism used by the act to require these collaborations is the requirement that the Secretary of Labor develop an annual plan for the national employment statistics program; the consultation with the states is intended to be a key input to the annual plan.

> **Box 3-1** Surveys Comprising the Common Core of Data
>
> *Public Elementary/Secondary School Universe Survey*
> - institutional characteristics, numbers of teachers, enrollment by grade, students participating in selected education programs, dropouts, and high school completers
>
> *Local Education Agency (School District) Universe Survey*
> - institutional characteristics, number of education staff, and number of students participating in selected education programs
>
> *Local Education Agency (School District) Universe Survey: Dropout and Completion Survey*
> - number of dropouts from each of grades 7 through 12, and the numbers of high school diploma recipients and other high school completers
>
> *State Nonfiscal Public Elementary/Secondary Education Survey*
> - state-level counts of students, teachers, other staff, and high school completers
>
> *National Public Education Financial Survey*
> - state-level collection of revenues and expenditures
>
> SOURCE: Adapted from workshop presentation by White and descriptive material at the CCD website (http://nces.ed.gov/ccd/index.asp [April 2009]).

### 3–C.2 The Common Core of Data

Andrew White (National Center for Education Statistics, NCES) described the CCD, the U.S. Department of Education's primary database on public elementary and secondary education in the United States. Administered by NCES, the program annually collects fiscal and nonfiscal data about all public schools (approximately 96,000), public school districts (approximately 18,000), and the 50 states, the District of Columbia, Department of Defense Schools, and those in outlying areas.

The CCD comprises five surveys sent from NCES to state education agencies (typically a state department of education) and completed mostly by using administrative data already maintained by the agency; the surveys are listed in Box 3-1. White noted that completing some data items requires the agency to contact local school districts, who in turn may contact individual schools. The state agency then compiles the data from all levels into prescribed formats and transmits them to NCES. The data gathered from the surveys fall into three categories:

1. general descriptive information on schools and school districts, including name, address, phone number, and type of locale;

2. data on students and staff, including selected demographic characteristics; and

3. fiscal data that include revenue and current expenditures.

White said NCES began entering into cooperative partnerships in 1985 in an attempt to improve CCD data. Of note was a contract with the Council of Chief State School Officers (CCSSO) to examine the completeness and comparability of data reported to the CCD, as well as to discuss ways to expanded its content and establish common definitions.

The Hawkins-Stafford Elementary and Secondary Education Improvement Amendments of 1988 (P.L. 100-297) authorized NCES to establish a formal federal-state cooperative system. The goal of this system was to "produce and maintain, with the cooperation of the States, comparable and uniform educational information and data that are useful for policymaking at the Federal, State, and local level" (Hoffman, 2004:x).

To implement and support the new cooperative system, NCES, with assistance from CCSSO, formed the National Forum on Education Statistics (NFES) in 1989 with a mission (Hoffman, 2004:xii):

> To develop and propose, cooperatively, a national education data agenda and model(s) for a national data system that will meet the needs of education policy makers and program planners in the decade and beyond; To inform Federal, State, and local decision makers on the goals and progress of this cooperative education statistics system; To provide an arena in which Federal, State, and local education interests can identify, debate, mediate, and where appropriate, recommend action on education policy, issues, emerging needs, and technological innovation salient to the improvement of education data comparability, uniformity, timeliness, and accuracy at the national level.

The NFES also adopted the role of "provid[ing] direction for research and evaluation" and "bring[ing] to the attention of relevant parties such matters as may contribute to the accomplishment of this mission" (Hoffman, 2004:xii). Chief state school officers, federal program heads, and directors of professional associations with an interest in education statistics were asked to appoint liaisons who would represent their various institutions. The NFES formalized its goals, objectives, functions, organizational structure, and operations in January 1990 with the adoption of a Policies and Procedures Manual. In 1996, the NFES expanded its membership by adding one local education agency representative from each state, to be appointed by the chief state school officer.

To achieve its mission, the NFES holds regular meetings, including standing committees that address specific issues, and produces a number of reports, including a series of "best practice" guides on a wide range of data-related topics. NCES funds state participation in NFES activities and publishes and disseminates definitions and guides from NFES.

In 2003, the department launched the Education Data Exchange Network Submission System (EDEN) to provide a common system by which state education agencies could transmit their administrative data. Data are transmitted by the states to meet the data requirements of annual and final grant reporting, specific program mandates, and the Government Performance and Results Act. In addition, the EDEN Survey Tool was established to allow transmission of additional data, such as the Civil Rights Data Collection and the Indian Education Formula Grant Program Application for Funds.

In 2006, the Department of Education launched a more overarching system called EDFacts which is a central portal for performance and accountability data reporting, including nonfiscal CCD data.[5] White noted that implementation of EDFacts has "done a little damage" to the October 1 reporting deadline. In particular, states that had highly developed data systems in place prior to EDFacts have had a difficult time converting to the new format and its definitions. The department has provided some funding to states to help them enter the EDEN/EDFacts system. Since January 2007, reporting of these data using EDFacts is mandatory (with a 2-year transition period).

The establishment of the Statewide Longitudinal Data System Grant Program has also provided an opportunity for states to apply for grants from between $500,000 and $6 million to develop and implement longitudinal data systems. The grants provide funding for 3-year cycles. Participation has grown substantially every year since the program's inception in 2005. Many states have been awarded their second 3-year grant, and only eight states have not participated.

---

[5] For more information on EDEN and EDFacts, see http://www.ed.gov/about/inits/ed/edfacts/overview.html (April 2009).

# – 4 –

# Methodological Issues and the 2003 Revision of Standard Instruments

THE CONTENTS OF THE RECORDS of vital events in the United States have not been stagnant over time. In its role of coordinating collection at the national level, the National Center for Health Statistics (NCHS) periodically revises the recommended U.S. standard certificates of birth, death, marriage, and divorce, and related records. However, in the Vital Statistics Cooperative Program, use of these standard certificates is not legally binding on the states and registration areas, although consistency in the data collection instrument certainly affects the quality of resulting data.

In 1998, NCHS and the vital statistics stakeholders began the most recent revision of the standard instruments, resulting in the 2003 release of new instruments. The 2003 revision marked the 12th revision of the birth certificate and the 11th revision of the death certificate, and both marked the first revision of the standard certificates since 1989. A particularly important feature of the new certificates was revision of items for collecting information on race and Hispanic origin, in compliance with new U.S. Office of Management and Budget (OMB) regulations to permit the reporting of multiple-race categories. However, for a variety of reasons, the vital registration areas have been slow to adopt the new form of the certificates, leading to significant methodological challenges in recent years. The patchwork of adopting and nonadopting registration areas has forced attention to how data on race can and should be tabulated in vital statistics and used in such important applications as population estimates; the uneven implementation of other new demographic and health data items on the standard certificates has also limited capacity to use vital statistics data for research purposes.

This chapter describes the 2003 revised instrument in some additional detail, consistent with the material presented at the workshop. Three workshop presenters from the NCHS staff and one Census Bureau speaker commented on the methodological changes inherent in the 2003 revision and the types of analyses made possible by new variables added in the revision. Specifically, the presentations focused on the complications involved in working with the data in the current situation in which implementation of the revised certificate is uneven and, hence, states report information in varying formats.

## 4–A THE 2003 REVISIONS

NCHS convened an expert panel in 1998, consisting principally of state vital registration officials, as well as representatives from relevant user organizations, to begin the process of evaluating the content of the existing (1989 revision) birth and death certificates and recommend changes. The panel (Division of Vital Statistics, 2000) developed its final recommendations in 1999 and directed that NCHS test redesigned instruments; the resulting documents became the 2003 standard certificates, and they are reproduced in Appendix D.

The major changes to the standard certificates are described in brief in Box 4-1. As Jennifer Madans (NCHS) observed in her workshop presentation, the 2003 revision continued a long-term push to make the vital records a platform for collecting a variety of public health data items in order to meet real public health needs. The revised 2003 birth certificate now includes some 60 data items, providing extensive information on pregnancy, labor and delivery, infant health, and maternal health factors; the 2003 round specifically added queries on risk factors (smoking) and method of delivery.

Arguably the most significant change made in both instruments was described in more detail by workshop presenter James Weed (former deputy director, Division of Vital Statistics, NCHS; retired): modification of the questions on race and Hispanic origin to reflect new standards promulgated by OMB in 1997. As Weed summarized, the standards (U.S. Office of Management and Budget, 1997):

- established a minimum set of race categories that were made mandatory for statistical data collections: American Indian or Alaska Native; Asian; black or African American; Native Hawaiian or Other Pacific Islander; and white;
- defined a minimum set of categories for collection of Hispanic or ethnic origin: "Hispanic or Latino" or "not Hispanic or Latino"; and

**Box 4-1** Major Changes to the U.S. Standard Certificates for Vital Events, 2003 Revision

**U.S. Standard Certificate of Live Birth**
- *New items*
  - Fertility therapy
  - Use of WIC (Special Supplemental Nutrition Program for Women, Infants, and Children) funds to obtain food during pregnancy
  - Infections during pregnancy
  - Maternal morbidity
  - Breast feeding
  - Principal source of payment for the delivery
- *Modified items*
  - Mother's and father's race, for compliance with U.S. Office of Management and Budget (OMB) standards
  - Mother's and father's education, to record highest degree attained by both
  - Level of smoking before and during pregnancy
  - Method of delivery question, to include trial of labor prior to cesarean delivery and other categories
  - Pre-pregnancy weight of mother and weight and height of mother at delivery
  - Congenital anomalies
- *Related documentation*
  - "Worksheet" to be filled by mother, giving self-response to such items as personal characteristics and program participation
  - "Worksheet" to be filled by facility (and accompanying guide), based on medical records, covering items such as birth weight and method of delivery
  - Detailed specifications and instructions for every element in the electronic birth certificate system
  - The U.S. Standard Report of Fetal Death was modified for conformity with revised items on the new birth certificate; the record also queries for other significant causes of death in addition to a single initiating cause and adds items on autopsy or histological placental examination

**U.S. Standard Certificate of Death**
- *New items*
  - Maternal mortality (pregnancy status at time of death)
  - Decedent's role in the event of death due to transportation injury (e.g., passenger, driver)
  - Tobacco use and contribution to death
- *Modified items*
  - Decedent's race, for compliance with OMB standards
  - Decedent's education, to record highest degree attained
  - Decedent's marital status, to distinguish "married" category from "married, but separated"
  - Place of death, to include hospice facility
- *Related documentation*
  - Revision of funeral director's handbook for completing death certificate
  - Revision of physicians' and medical examiner/coroners' handbooks to focus on accurate collection of cause-of-death information
  - Detailed specifications and instructions for every element in the electronic birth certificate system
  - Certificate includes separate instructions specifically for funeral director and for person completing medical certification portion

SOURCES: Workshop presentations summarized in this chapter; Division of Vital Statistics (2002a,b).

- indicated a preference for asking the two questions (race and Hispanic origin) as separate items, with the Hispanic origin question preceding the race question.

The five base racial categories differed from previous standards by splitting the previous Asian and Pacific Islander category into two and by intending to be inclusive (i.e., there was no "Other" option). However, the significant change in the 1997 OMB standards was to permit multiple selections rather than just one, allowing respondents to self-identify with more than one race category. Weed commented that the directive also requested that agencies show as much multiple-race detail in its tabulations as possible, subject to data quality and confidentiality standards, and that agencies not ask respondents indicating more than one race to pick one as a "main" or "primary" identification.

Many vital records jurisdictions have been slow to fully implement the 2003 revised certificates. A survey of 52 of the 57 jurisdictions by Friedman (2007:Table 1) suggests that the 1989 revision of the standard birth and death certificates won rapid acceptance by the local authorities.[1] Specifically, Friedman (2007:Table 1) found that 50 jurisdictions implemented the new standard birth certificate in 1989 with the other two areas complying within 2 years; 48 jurisdictions supplied data using the new death certificate format in 1989, three followed in 1990, and only one took longer to implement (in 1997). By comparison, only two and five states used the 2003 revised birth and death certificate formats in 2003, respectively. As of the time of the Friedman (2007) survey—more than 3 years later—27 and 26 jurisdictions had not yet implemented the new birth and death certificates, respectively.

Several workshop speakers presented maps depicting the current level of implementation of the standard certificates; this information is summarized in Table 4-1. Speaker Robert Anderson (NCHS) added that six registration areas are planning on implementing the revised death certificate in 2009 and another seven in 2010.

## 4–B  RACE AND ETHNICITY

Weed noted that, at the time of the 1997 establishment of OMB's revised standard for race and Hispanic origin questions, agencies were required to implement the new standards by January 1, 2003 (with some allowance for a "bridging" period between single-race and multiple-race reporting—discussed further, below). In particular, the Census Bureau implemented the new categories and "mark all that apply" approach to the questionnaires used in the 2000 census. However, Weed commented that the vital statistics

---

[1] The Friedman (2007) analysis was mentioned and commended by Harry Rosenberg in his workshop presentation.

**Table 4-1** Adoption of 2003 Revised Certificates and Multiple-Race Reporting for Births and Deaths, by State, 2005

| State | Using 2003 Revised Certificate? | | Multiple-Race Reporting Allowed? | | State | Using 2003 Revised Certificate? | | Multiple-Race Reporting Allowed? | |
|---|---|---|---|---|---|---|---|---|---|
| | Births | Deaths | Births | Deaths | | Births | Deaths | Births | Deaths |
| Alabama | ○ | ○ | ○ | ○ | Montana | ● | ● | ○ | ● |
| Alaska | ○ | ○ | ○ | ○ | Nebraska | ● | ● | ● | ○ |
| Arizona | ○ | ○ | ○ | ○ | Nevada | ○ | ● | ● | ● |
| Arkansas | ○ | ● | ○ | ○ | New Hampshire | ● | ● | ● | ● |
| California | ● | ○ | ● | ● | New Jersey | ● | ○ | ○ | ○ |
| Colorado | ○ | ● | ● | ○ | New Mexico | ● | ● | ● | ● |
| Connecticut | ○ | ● | ● | ● | New York | ○ | ○ | ○ | ○ |
| Delaware | ○ | ◐ | ● | ◐ | North Carolina | ● | ● | ● | ● |
| District of Columbia | ○ | ● | ● | ● | North Dakota | ● | ● | ● | ● |
| Florida | ● | ● | ● | ● | Ohio | ○ | ○ | ● | ● |
| Georgia | ○ | ○ | ○ | ○ | Oklahoma | ● | ● | ● | ● |
| Hawaii | ○ | ○ | ● | ● | Oregon | ● | ● | ● | ● |
| Idaho | ● | ○ | ● | ○ | Pennsylvania | ● | ● | ● | ● |
| Illinois | ○ | ○ | ○ | ● | Rhode Island | ○ | ○ | ● | ○ |
| Indiana | ○ | ● | ○ | ● | South Carolina | ● | ● | ● | ○ |
| Iowa | ○ | ○ | ● | ● | South Dakota | ● | ● | ● | ● |
| Kansas | ● | ● | ● | ○ | Tennessee | ● | ● | ● | ● |
| Kentucky | ● | ○ | ● | ● | Texas | ● | ○ | ● | ● |
| Louisiana | ○ | ○ | ○ | ○ | Utah | ◐ | ● | ◐ | ○ |
| Maine | ○ | ○ | ● | ● | Vermont | ○ | ○ | ○ | ○ |
| Maryland | ○ | ● | ● | ● | Virginia | ○ | ○ | ● | ○ |
| Massachusetts | ○ | ○ | ○ | ○ | Washington | ● | ● | ● | ● |
| Michigan | ○ | ● | ◐ | ● | West Virginia | ● | ● | ○ | ● |
| Minnesota | ○ | ● | ● | ● | Wisconsin | ○ | ○ | ○ | ○ |
| Mississippi | ○ | ○ | ○ | ○ | Wyoming | ● | ● | ○ | ● |
| Missouri | ○ | ○ | ○ | ○ | Total ● | 12 | 17 | 17 | 21 |

NOTES: ●, Yes; ○, No; ◐, Partial (for only part of data year 2005 or only for selected facilities). "New York" is New York state less New York City, which reports separately. Reporting status in Puerto Rico and the territories not shown in this table.

SOURCES: Data on births from Martin et al. (2007:89, 93); data on deaths, from Kung et al. (2008:105, 107).

program was granted a variance or exception to this deadline, because of the difficulty in achieving simultaneous compliance by all the vital records reporting agencies. When changes were being made to the standard birth and death certificates in 1999–2000 (what would eventually become the 2003 revisions), Weed said that NCHS decided that its best approach would be to try to emulate the phrasing and construction of the census questions to the greatest extent possible. This would not only permit (eventual) comparability between the two resulting data sources, but would also allow the use of similar coding and editing processes (e.g., to resolve inconsistent or redundant write-in responses) by the two agencies (Division of Vital Statistics, 2004). The certificate revisions followed the same category breakdowns as the 2000 census questionnaire for the Hispanic origin question but differed slightly from the census instrument by more detailed splitting of the Asian and Pacific Islander categories (the census permitted only one space to write in either an "Other Asian" or "Other Pacific Islander" affiliation).

At the time of the workshop in 2008, the most recent issues of final vital statistics covered data year 2005 (Kung et al., 2008; Martin et al., 2007). For that year, Weed said that multiple-race reporting for deaths had been implemented in 22 of the 57 vital record jurisdictions and in 17 jurisdictions for births.[2] For those jurisdictions, multiple-race reporting for decedents (information from next of kin) was indicated on 0.4 percent of the records, more frequently for younger decedents (2.4 percent of those under 25 years of age) than for older decedents (0.3 percent of those over 64). Higher levels of multiple-race reporting are indicated in the birth records for the available states—1.5 percent, ranging geographically from 0.4 percent (Texas) to 36.6 percent (Hawaii).

Weed noted that NCHS had received multiple-race data for deaths from 30 jurisdictions in 2007 and anticipates compliance by 40 jurisdictions in 2008; for births, 30 jurisdictions reported multiple-race data in 2007, expected to grow to 47 in 2008. Weed observed that, until all states consistently report multiple-race data in the same manner, the vital records suffer from two fundamental compatibility problems. First, they are not compatible between jurisdictions, which may differ in their reporting schemes. In particular, one approach that has been used in the past (and may still be in use in some localities) for coding multiple-race entries from a birth or death

---

[2] According to Weed's presentation, multiple-race reporting in 2005 was done for *both births and deaths* in California, Florida, Hawaii, Idaho, Kansas, Minnesota, Nebraska, New Hampshire, New York state, South Carolina, Utah, and Washington. Multiple races for *deaths only* were reported in Connecticut, Maine, Michigan, Montana, New Jersey, New York City, Oklahoma, South Dakota, Wisconsin, and Wyoming (with the District of Columbia reporting for part of the year). Reporting for *births only* occurred in Kentucky, Ohio, Pennsylvania, Tennessee, and Texas (with Michigan and Vermont reporting for part of the year or only for selected facilities).

certificate is to record only the first one mentioned. Second, the state-level records are not compatible with the data collected in the decennial census (or the new American Community Survey) or with the Census Bureau's intercensal population estimates.

Relatively little is yet known about the characteristics of persons who report multiple race categories in comparison with those who report a single race, the internal cognitive weightings that may go into such determinations (e.g., consistent reporting of American Indian or Native Hawaiian ancestry), or whether the multiple categories accurately capture an individual's sense of racial and ethnic identity.

### 4–B.1 Bridging Single-Race and Multiple-Race Data at NCHS

Weed mentioned some insights that have been derived from analysis of data from the National Health Interview Survey (NHIS) and work on "bridging" generally: using statistical modeling of multiple-race responses to attempt to identify the one single response that an individual would have reported under the old single-race standard.

The NHIS (fielded by NCHS) began permitting multiple responses to race questions in 1982; respondents could choose up to two categories through 1996 and as many as five categories beginning in 1997. Those respondents checking more than one category were also prompted to name their primary or main race affiliation—the one they say would best represent them. With the promulgation of the new OMB standards, NCHS won a special variance for the NHIS to continue to ask the follow-up question on a preferred single-race category. The resulting data have played a key role in developing methodology for bridging multiple-race responses to single-race categories: using the percent distribution of the "preferred"/single-race categories for each multiple-race combination to proportionally allocate persons to a single race. Weed presented three specific examples from 4 years (1997–2000) of the NCHS data:

- For persons reporting the two-race combination of black and white, 45.4 percent identified black as the single-race category that best represents them. Persons indicating white as a primary race category and those indicating no preferred single race were about evenly divided, at 26.9 and 27.7 percent, respectively.

- For persons reporting the two-race combination of American Indian/Alaskan Native and white, the distribution was more lopsided: 74 percent identified white as their primary race affiliation, 21.2 percent American Indian or Alaskan Native, and 4.8 percent no primary race.

- For persons indicating the three-race combination of white, black, and American Indian/Alaskan Native, the modal response was to indicate no primary race affiliation (57 percent); 27.6 percent indicated black as the primary preference and 8.5 percent and 6.9 percent chose white and American Indian, respectively.

For multiple-race combinations with enough reports in the 4 years of NHIS data to support further analysis, NCHS developed its algorithm for bridging or allocating a single-race response using multinomial logistic regression. The regression models included county-level covariates (urbanization level, log percentage of single main-race reporting in the county, percentage of multiple-race reporting in the county); NCHS has shared its bridging algorithm with the Census Bureau to use in analyzing and developing its population estimates. In Weed's assessment, the algorithm provides reasonably consistent numerators and denominators for such estimation purposes until all birth and death counts are available in the new multiple-race format. NCHS staff also contributed to an analysis of the performance of the bridging algorithm on the multiple-race entries to the 2000 census; see Ingram et al. (2003).

Going forward, Weed suggested that vital statistics data would be an important proving ground for better understanding the characteristics of persons who report multiple-race affiliations. In large part, this is because of sheer numbers: along with the decennial census and the American Community Survey, data on births and deaths are one of the data sources in which multiple-race combinations are reported in sufficient quantities to support detailed analysis. Weed noted that NCHS colleagues had already begun some studies in this regard, including the analysis by Hamilton and Ventura (2007) of births to mothers who report a single race or multiple-race affiliations in California and selected other states.

### 4–B.2  Bridging Single-Race and Multiple-Race Data at the Census Bureau

In her workshop remarks, Victoria Velkoff (U.S. Census Bureau) commented on the challenges of working with the new multiple-race categories in the Census Bureau's population estimates program and reconciling them with the different race measurement in the vital statistics program. Velkoff acknowledged that the need to bridge from the "old" race categories in vital statistics records to new categories complicates the estimation process.

Velkoff indicated that most of the birth data that the Census Bureau receives directly from the states do not use the current race categories. Hence, the Census Bureau uses models to develop the birth component of the population estimates using the race information for parents included on the birth certificate and the distribution of family composition for multiracial families from the 2000 census. For death data, Velkoff noted the fundamental challenge that the decedent race information on death certificates is, by default, a proxy report rather than a self-report. Although the race data are supposed to be provided by a family member or next of kin, it sometimes falls to a funeral director or medical examiner to fill in that category, resulting in inconsistencies. These inconsistencies (including, perhaps, incomplete reporting of American Indian heritage or multiracial combinations) carry over to affect the intercensal population estimates.

For its adjustment to the base population in constructing estimates, Velkoff indicated that the Census Bureau is also still grappling with another feature of race reporting in the 2000 census. The 2000 census questionnaire permitted respondents to choose multiple options from the five race categories defined in the 1997 OMB standards but also included a sixth option, "some other race." For its population estimates, Velkoff said that the Census Bureau distributes the "some other race" responses into the standard categories—a second level of bridging—so that the estimates do not include tabulations for "some other race."[3]

The problem of working with varying race categories has been more acute in the area of population projections because, historically, the projections have used a coarser categorization. Fred Hollmann (Census Bureau) said that, prior to 1986, the Census Bureau performed projections for only three categories—white, black, and other races—with "other" implied as a residual. In 1986, the Bureau performed a special projection of the Hispanic population through 2080, using vital statistics and birth certificate data for Hispanic births in 22 states. At that time, the Census Bureau deemed the death certificate reporting of Hispanic origin to be insufficient in detail, and so it relied instead on life tables developed for the state of California (augmented by Medicare data). By 1993, the population projection program began the practice of crossing Hispanic origin with four race categories. As of the 2004 release of interim projections, the Census Bureau is using the same 31 race and multiple-race categories, crossed with Hispanic or not Hispanic, as the population estimates program. Its products use what has been called a "min-max" approach, defining only two values for each of the five race categories: those who report the race as the only race and those who report the race either alone or in combination with (any number of other) categories.

---

[3]The Census Bureau tested alternatives to the race and Hispanic origins in its 2003 and 2005 tests that omitted "some other race" as a response option. However—presumably concerned that respondents who strongly self-identify with their Hispanic origin as a race (rather than a separate ethnicity category) choose "some other race" as the most fitting option—congressional appropriators have forbidden such changes. First enacted in the Consolidated Appropriations Act of 2005 (P.L. 108-447) and repeated in some subsequent reports, appropriators directed that "none of the funds provided in this or any other Act for any fiscal year may be used for the collection of Census data on race identification that does not include 'some other race' as a [category]."

Internally, though, the Bureau's models of fertility and mortality rates are still principally based on three main categories: white, non-Hispanic black, and other. Hollmann suggested that the Census Bureau plans to revisit its projection strategy after all the states adopt the 2003 revised standards for birth and death certificates and report race in a consistent fashion.

Hollmann said that NCHS's switch in 1989 from tabulating vital statistics on birth—from the (imputed) race of the child to the race of the mother—was a significant difference for the population projections program. Initially, the Census Bureau tried a "workaround" of adapting race of mother to infer race of child using some information on racial characteristics of families from a fertility supplement to the Current Population Survey. However, that approach proved inconvenient and burdensome; accordingly, the Bureau elected to base its projections primarily on the race of the mother in determining the "age 0" population of newborns in each subsequent year. With the promulgation of the 1997 standards, the Bureau changed again to an imputation strategy, deriving a child's race from the race of both parents. However, Hollmann reported that the imputation process continues to be somewhat problematical (though arguably more so for the population estimates than for the projections) and an area of continued work.

## 4–C  FETAL DEATHS AND INFANT HEALTH RISK FACTORS

In the U.S. vital records system, reports of fetal deaths are completed separately from certificates of birth and death. The fetal death report contains additional questions on the cause and condition of the death, demographic and health information on the mother, record of previous prenatal care, and risk factors involved in the pregnancy. The Report of Fetal Death is periodically revised in the same manner as the Standard Certificates of Birth and Death; as shown in Box 4-1, the standard record was revised in 2003 to expand collection of cause-of-death data and include additional questions also added to the standard birth certificate.

Stephanie Ventura (chief, Reproductive Statistics Branch, NCHS) noted that the process of adoption of the 2003 revision of the U.S. Report of Fetal Death has been slower than that of the standard birth certificate (as shown in Table 4-1). By 2006, only 19 states had adopted the 2003 revision of the birth certificate; Ventura said that those states account for about half of all births in the country. NCHS anticipated that adoption of the new form in about eight additional states and New York City by 2008 would push that coverage to about two-thirds of all births.

The states that are using the 2003 format and questions are an incomplete set, but NCHS has generally found them (particularly a 12-state subset as of 2005) to be representative of the national population in terms of racial

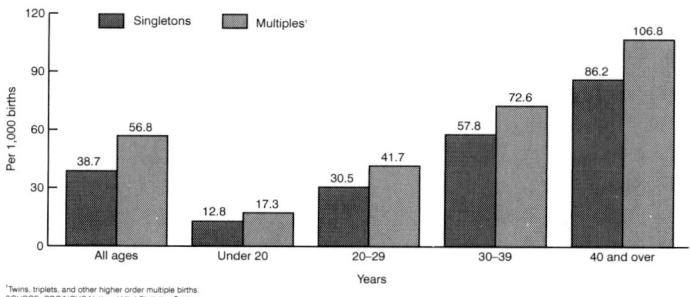

**Figure 4-1** Rates of gestational diabetes by age of mother and plurality, 12-state reporting area, 2005
SOURCE: Menacker and Martin (2008:Fig. 1).

and ethnic composition. Ventura summarized findings from recent NCHS reports that use the new additional information included on both the birth certificate and record of fetal death. In addition to restriction to the 12-state subset, Ventura discussed what NCHS terms "releasable" data—new checkbox items in existing data items that are already part of the national data set and that the states have authorized NCHS to release.

- *Diabetes*—Prior to the 2003 revision, only a single check box was available to indicate diabetes as a risk factor; the new certificate includes a distinction between preexisting and gestational diabetes. Figure 4-1 illustrates reported gestational diabetes by the age of the mother and plurality of birth (i.e., whether the birth is a singleton or a multiple delivery). The data suggest that the risk of gestational diabetes is elevated for older mothers, regardless of plurality.

- *Racial disparities*—Some important racial disparities can be found through analysis of birth certificate data. Figure 4-2 shows the receipt of surfactant replacement therapy by gestational age, a procedure to directly provide surfactant before the lungs are capable of producing it naturally, to make lung expansion and breathing easier. As is to be expected, the data show that the procedure is most frequently used for extremely preterm births for which it is most necessary to prevent respiratory distress syndrome. However, the data also suggest that white births are much more likely to receive this therapy than black or Hispanic births, regardless of gestational age. Similarly, Figure 4-3 illustrates receipt of steroids by the mother, prior to delivery, for fetal lung maturation. Again, the procedure is most frequent for very

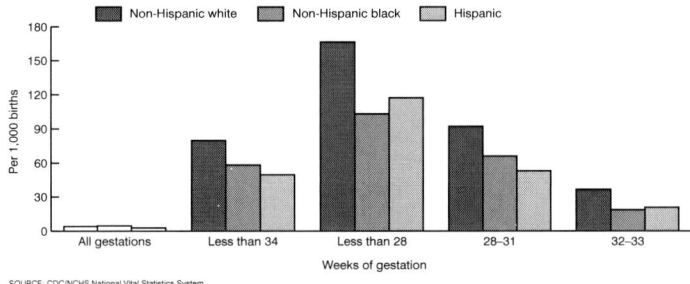

**Figure 4-2** Rate of surfactant therapy by gestational age and race and Hispanic origin of mother, 12-state reporting area, 2005

SOURCE: Menacker and Martin (2008:Fig. 6).

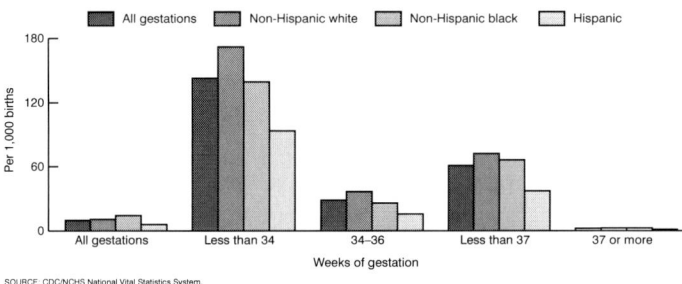

**Figure 4-3** Steroids for fetal lung maturation received by the mother prior to delivery, by gestational age and race and Hispanic origin, 12-state reporting area, 2005

SOURCE: Menacker and Martin (2008:Fig. 4).

preterm births, but white births are much more likely to receive this therapy regardless of gestational age.

- *Admission to neonatal intensive care units*—Analysis of distribution of admissions (for these purposes, restricted to singleton births) to neonatal intensive care units (NICUs) by gestational age shows an unexpected trend. The natural expectation is that preterm births would dominate those admissions, requiring the additional specialized care. However, almost half (47.5 percent) of those admissions were term or full term (37 or more weeks) at birth. Ventura noted that the factors leading to NICU admission are important to understand in planning for health care services, and NICU admission was not included as a specific factor on the standard birth certificate until the 2003 revision.

In addition to its work with the new data items, Ventura noted that NCHS is pursuing research and improvement efforts in its traditional birth and infant health data. Particular effort is being put toward improving the data quality of the fetal death files. In addition, the generation of a linked birth and infant death data set (described in Box 1-1) has been a particular asset for research. Ventura said that the 2006 data suggest that about one in eight births in the United States is preterm, a percentage that has been growing and that heightens the importance of studying preterm-related causes of infant deaths and the factors that contribute to successful deliveries.

Still, Ventura said that NCHS is beginning to focus predominantly on the revised data, even though transition to the new certificates is still in progress. Part of NCHS's strategy in this regard is continued improvements in data availability. Birth and perinatal data are available for online tabulation and mapping at NCHS's VitalStats site (http://www.cdc.gov/nchs/VitalStats.htm), and more data sets and documentation are directly downloadable from the NCHS site as well (births from 1968 through 2005; linked birth-infant death files from 1983 through 2004). However, Ventura emphasized that the major goal for the revision is not simply adding new data items, but also improving the data quality of the overall vital statistics series. Although resources for evaluation have been limited for a number of years, Ventura argued that the addition of the new data makes a program of data validation essential; it is critically important to assess the consistency of reporting in new and old data items and their comparability. She said that NCHS would like to be able to more regularly initiate special projects with individual states or groups of states to focus on data collection and data quality issues.

## 4–D  MORTALITY AND CAUSES OF DEATH

Robert Anderson (chief, Mortality Statistics Branch, NCHS) commented on the major methodological issues faced by NCHS with regard to the mortality data, in particular the handling of cause-of-death data. In contrast with birth data, fewer states have adopted the 2003 revised standard certificates for death; hence, Anderson noted that NCHS is not quite to the point that it is with the birth data in terms of transitioning to "revised" data.

Anderson said that the timeliness of release of vital statistics mortality data has been reduced not only by the slowness of adoption of the revised certificates, but also by the ongoing challenge of implementing the 10th revision of the International Classification of Diseases by local authorities. He said that the adoption of electronic death registration systems has been slower and more difficult than electronic birth registration systems.

These challenges to timeliness have manifested themselves in the lags between data collection and publication of final national-level statistics. Anderson presented data on the past 12 years of data releases, showing the times between the end of the data year and the release of estimates, first as a preliminary data set and then as a final data release. The data showed increased lag times—particularly in the time until production of the preliminary estimates—since 1999. From 1995 to 2005, including a projection for release of 2006 data, the mean time to preliminary release was 13.8 months; the average time to release of final data was 21.5 months.

Anderson said that electronic death registration systems promise to greatly improve the timeliness of mortality data, after the initial growing pains of implementation. As an example, he noted that New Hampshire was the first state to implement such a system and is now able to provide the state epidemiologist with death certificate data, including cause of death, within 2–3 days from the date of death. A map displayed by Anderson suggested that electronic death registration systems were functioning in 22 states, the District of Columbia, and New York City; 8 states were said to be in development as of January 2008 and another 8 in planning or requirements stages. The presence of an electronic registration system in a state does not necessarily mean that it has complete coverage, but that the system has at least begun to be used.

Cause-of-death certification by physicians is an area of concern for the quality and consistency of mortality data, both in general and with specific regard to the development of electronic death registration systems. Anderson said that physicians are not always fully aware of the way in which death certificate data are compiled at the national level and used for public health and resource allocation purposes. For example, "cardiac arrest" is still listed as the sole cause of death on about 12,000 certificates each year (though that number has decreased somewhat over time): from the data standpoint, this is undesirable because the terminal event of cardiac arrest provides no information on the mechanism of death or the factors that contributed to death. Disseminating training materials to the local level—including the development of online training tutorials—is an important step in correcting these problems. Anderson also said that states are experiencing some problems in getting physicians to accept the concept of electronic certification, which is obviously an important part of a fully fledged electronic death registration system.

Regarding the new data items added in the 2003 certificate revision, Anderson noted that the change in the education variable—from years of education to highest degree attained—has been somewhat problematic. The variable is important as a summary measure of socioeconomic status, essentially the only such measure available on the mortality data files. Because no way to exactly bridge between the two formats is available, NCHS is currently putting both education items on the data file (depending on the reporting area) so that users can consider both.

The perennial problem of death certificate data is that it is necessarily proxy-reported information; in particular, funeral directors are typically responsible for the sociodemographic information on the certificate. Though these data items, such as education, race, and Hispanic origin, are supposed to be obtained from a knowledgeable informant such as a family member, data quality and consistency problems arise because the data items are sometimes filled by observation. This phenomenon is thought to produce underreporting of Hispanic ethnicity and American Indian and Asian and Pacific Islander races; Anderson also said that it may be responsible for overreporting of high school completion. Anderson said that NCHS is hoping to conduct a national longitudinal mortality study, linking death certificate information with the Current Population Survey or other source, to try to better assess the difference between self-report and proxy-report items on death certificates.

# – 5 –

# Options for a 21st Century Vital Statistics Program

IN THE FINAL SESSION OF THE WORKSHOP, Jennifer Madans (National Center for Health Statistics, NCHS) outlined a variety of conceptual challenges and possible future directions for the Vital Statistics Cooperative Program (VSCP). These directions—and vital statistics as a whole—were then briefly addressed by each member of a panel of discussants; the timing of the workshop left some time, albeit not substantial, for general floor discussion.

Madans noted that probably the biggest challenge in the VSCP is that neither NCHS nor the vital registration areas have, or can have, control over the original source of the data. The individual physicians, funeral directors, and others who complete the information on certificates do not work for the state registrars and, hence, the registrars and NCHS have only limited means to affect the quality of the information that is collected. The 2003 revision of the standard certificates and related materials sought to make concepts easier and to provide better guidance (e.g., through the development of separate worksheets), but, ultimately, there is no direct control over the interaction between the data providers and the collector. At the national level, NCHS can promulgate standards but cannot mandate them.

Going forward, Madans suggested that some attention is needed on fundamental questions, such as: Is the current system of collecting a wide variety of data items on the birth certificate the best or most efficient method for getting that information? As Steven Schwartz and Ed Hunter observed in their presentations, birth certificates are administrative and legal documents with well-defined civil registration and identification purposes. Madans noted that early certificates were more focused on those needs but, over time, the number of items on the certificates grew. There were good reasons to add more public health data items to meet real public needs. Madans noted that the U.S. certificate may be unusual in the number of data items; by comparison, for example, Canadian birth certificates stick closely to the civil registration model and are limited to the most basic items (name, sex, and date and place of birth). The question is whether the U.S. certificate (with its 60 data items in the 2003 revision; see Appendix D) is now too large: What is known about the usefulness and quality of all that information, and what are the real cost implications of adding another item to the collection? To be clear, Madans said, this is not to say that the data items are unimportant by any means—just to question whether the birth certificate is the best vehicle for getting high-quality data efficiently.

More generally, Madans spoke of four issues for the future of vital statistics that all need to be dealt with, simultaneously, in choosing a future direction:

- *Infrastructure:* Current investments at the state and local levels have been geared to the electronic infrastructure of data processing and collection: indeed, the objective in the general category of infrastructure is to use information technology to achieve a faster, more efficient structure. Among the developments to date have been work on electronic verification of vital events, development of transmission standards, and creation of web-based systems for local practitioners to use in completing certificate information. Madans noted that these are all important advances, and advances that have to continue, but it has to be recognized that the "payoff" of these developments to date has not been as great as hoped. The phased nature of the implementations has been disruptive for trend analysis and overall comparability; Madans commented that it is possible that there might have actually been more rapid adoption of the 2003 standard birth certificate among the states if not for the difficulty of retooling electronic birth certificates. Madans also referred to statements made earlier in the workshop in commenting that infrastructure alone cannot solve all problems; even with the fastest and most efficient infrastructure, vital statistics are only as fast as the slowest state.

- *Content:* Picking up on her earlier theme—and taking care to note that suggesting changes to data content is often unpopular but nonetheless necessary—Madans said that the system needs to carefully consider what items are on the birth and death certificates and why. Is there some limited, core set of public health statistical data items that are truly essential? Are there additional items beyond that core for which the vital record is the only possible source, the most accurate possible source, or the most efficient source? If noncore items are placed on the certificates, is there a way of obtaining help from other agencies to pay for them?

- *Short-term considerations:* What are the most pressing current needs in the system and in terms of public health knowledge? Madans said that it has to be understood as a basic premise that current funding for the VSCP cannot sustain the current system and content into the future. Against this backdrop, the notion of a core of content and objectives to meet current needs becomes more critical. The current system is already making some sacrifices, in terms of timeliness and (to a lesser extent) data quality, in directing resources to future infrastructure investments: Can the system make better choices about those sacrifices to meet both short- and long-term needs?

- *Long-term considerations:* The goal of comprehensive electronic medical records is an elusive one. Madans commented that she was pessimistic about it being fully implemented in the near future, but she noted that it is eventually coming and that the VSCP needs to start taking it into account because it could fundamentally change the collection of vital statistics. If a lot of the information that is currently collected on vital records is maintained on the electronic medical record attached to each individual, then vital statistics becomes less a data collection problem than a record linkage activity. Looking ahead, it will also be important to reconcile the quality of the data from administrative and electronic medical records with the data from existing collection systems.

In sketching out some proposals for the future of vital statistics in light of these constraints, Madans emphasized that none of these proposals is desirable; all involve real pain, very difficult choices, and loss of data. She described three basic proposals:

1. NCHS could decide (with its stakeholders) that vital statistics are so critical that it would sacrifice all of its other data collections in order to fully fund and support the VSCP. The obvious difficulty with this approach is that there are large and vocal constituencies for other NCHS programs, such as the National Health and Nutrition Examination Survey (NHANES) and the National Health Interview Survey. Those constituencies would stridently argue the opposite: for example, the strong interest in biomarkers would suggest that NHANES should have primacy and everything else be terminated to fund it. Any major sacrifice of the data collection system is going to be extremely unpopular with a different

constituency—and for good reason, because these other data collection systems provide equally valuable information.

2. NCHS could completely revisit the cost structure of vital records and vital statistics. Madans said that it would help if funding for vital statistics had a more transparent cost model that could more readily convey the full costs, the different uses of different data items, and, accordingly, who should pay for them. These are not easy things to do because there are joint, overlapping uses. Just as NCHS and the states have worked to show the use and the quality of the data, a first step in a future system would be to have a fuller, more informed discussion of the costs of data item additions and deletions and the way in which the costs of the system are distributed.

3. NCHS could implement some combination of major cost-saving measures and reinvest those savings to develop the infrastructure. Such extensive cost-saving measures might include
    - switching to an effective 2-year cycle for national-level collection and reporting of vital statistics, alternating between birth records one year and death records the next;
    - collecting only a sample of records; and
    - reducing the content in the data files for all, or a large subset, of records, consistent with the idea of a core of essential data items, obtaining funding from other sources for any noncore items.

As with the overall options, Madans argued that each of these cost-cutting measures is bad in different respects and would have deleterious effects for many data users (including several of the applications described at the workshop). Still, the challenge is to consider the advantages and disadvantages of each option or combination of options.

In discussion, Kenneth Prewitt added that the NCHS Board of Scientific Counselors had offered the agency one overriding piece of advice on cost-cutting measures: that across-the-board diminution of resources (and, consequently, quality) across all NCHS programs is not a viable option. Tough, if unpalatable, choices are preferable to a protracted death by a thousand cuts.

Madans said that there is probably more agreement about what a future vital statistics program should look like, as a goal, than the means to get there. The goal for vital statistics should be a system that is based on both current and future needs; that is efficient but adequately funded, drawing from a variety of sources if necessary; that optimizes data quality and timeliness; that takes advantages of information technology developments but is not limited by technology; and that informs research and decision making. The vital statistics program must also always be thought of as integrated with other data collection systems on health.

In reaction to these proposals, and the workshop topics in general, several invited experts provided their reactions.

Howard Hogan (U.S. Census Bureau) prefaced his remarks by suggesting that a nation's vital statistics system is a reflection of its social concerns, an impression he first derived from his dissertation work on infant mortality rates. For example, different countries emphasize details on nationality and parentage but nearly all countries, historically, detail legitimate or illegitimate births as an indicator of social conditions. To this extent, Hogan said, Madans's suggestion that the birth certificate is dominated by data items on race and Hispanic origin and on public health issues actually says something about us as a nation. Moreover, Hogan argued that a nation's vital statistics system is indicative of its overall statistical development. That Sweden was able to issue fairly good and detailed birth statistics says something about the social and statistical organization of Sweden in 1790; that alarms have been raised about funding crises and lags in data releases in U.S. vital statistics since at least 1975 says something else. Hogan said that vital statistics are, fundamentally and emotionally, a building block of a national statistics system and permitting them to wither away would say a great deal about the nation's commitment to good statistics and the social problems they reflect.

Hogan suggested that the workshop discussion—like conventional usage over the decades—equates two things that are not quite the same. Though "vital statistics" have been taken to mean "statistics tabulated from a vital events registration system," it is possible to interpret "vital statistics" more broadly. Hogan noted that various workshop presentations had alluded to the fact that the United States does not have vital statistics on marriages and divorce; this is true if looking at data from registration systems but it is not true if data from different systems are considered. The Census Bureau's American Community Survey (ACS) includes data items on marriage and divorce that appear to provide reasonably reliable results, including number of times married and year of last marriage. Divorce data are somewhat more problematic because people say they are divorced when they have only separated or applied for divorce, but the data still provide useful information. Similarly, the ACS also queries for some items that correspond to birth certificate data—mother's age, race, Hispanic origin, and nativity—that compare fairly well with the registration data. Although the registration data include much more detail on medical matters, the ACS arguably provides a richer array of social and economic data linked to the same births, permitting different kinds of analyses. Other data systems, such as the National Health Interview Survey, overlap the classic "vital statistics" data in important respects.

Hogan said that his remarks are not meant to imply that the vital record system is unnecessary. To the contrary, although the surveys can provide data richness, they depend heavily on the "census" of events, i.e., vital statistics from the registration system, for calibration and benchmarking. Logically, deaths are an example of a vital event for which no survey is capable of providing good and detailed information. In terms of the suggestions for cost savings in the vital statistics systems, Hogan suggested beginning by consideration of the vital statistics system as a whole and, from that perspective, think of what kind of data is needed for analysis of births, marriage, and other events. On the basis of this analysis, ideas such as Madans' notion of a core of content need to be considered but, more importantly, the system needs to consider the ways in which the necessary information can be obtained from means other than the vital records—either independent surveys or surveys using the registration system as a sampling frame.

Garland Land (National Association of Public Health Statistics and Information Systems) agreed with several of Madans' comments but suggested that the outlook for the current VSCP might be too pessimistic. He said that the impression that the current system is very costly and inefficient and is, essentially, not viable financially is too harsh: the current cost of obtaining birth and death record data (less than $2–3 per event) probably compares favorably with data collection in national surveys. Furthermore, he questioned the impression that there are no controls over source data collection in the VSCP. He said that the states do have controls in that they regulate hospitals, funeral directors, and physicians, and the licensing process does give the states some leverage to address data quality concerns. Similarly, he said that NCHS has the ability to "control" the states through the contracts and purchase orders that underlie the VSCP; NCHS has the ability to effectively mandate that data be collected in a certain way or a state will not be paid. He acknowledged that the issues are more complex than his summary allowed, but concluded that there are controls at both the national and state levels that can be used much more efficiently than they are at present.

In terms of data content and quality, Land said that the 2003 revision process did go somewhat awry in the major expansion of data items: in meeting demands for research, Land suggested that the result was a cumbersome birth certificate. Moreover, the additions were not grounded in a full evaluation of the items already on the certificate. He said that the individual data items vary widely in terms of their quality and require an item-by-item assessment; the current system lacks a means of removing underperforming items.

Land found the argument for a more transparent cost model uncompelling because of developments in recent decades. The cost models developed in the early 1980s were effective in attaching prices to parts of the VSCP system and were used to form the cost-sharing formula that was used for many years. That work suggested the appropriateness of a 50-50 model, under which the federal government and the states would split the costs. Land asserted that the problem is that the federal government never met that level; the largest federal share over the years has been about 30 percent, and Land estimated that the federal share is currently down to only about 6 percent of costs.

Land challenged the cost-saving options currently on the table as suffering from the same fundamental flaw: data collection may be staggered or data content reduced, which may produce some savings at the national level, but the states will still have to collect 100 percent of the data for the birth and death records. If states receive less money to collect the same amount of records, timeliness and quality are likely to suffer; sooner or later, similar to Schwartz's description of New York City's experience when funds for abortion reporting were eliminated (see Section 3–A in Chapter 3), the states and localities will be unable to afford coding clerks, trainers, or other support personnel, and data quality will decline.

Land expressed particular concern that so much is made of the small and steady or decreasing pool of resources available to NCHS as the sole possible source of funding for the VSCP. He said that there are major partners—for instance, the Social Security Administration and the Census Bureau—who might be looked to as sources of funding rather than only as receivers or users of the statistics. Land argued that the same kind of "marketing" of the data that is done at the state level is not adequately done at the national level: for instance, states have been active in finding new opportunities for use of the National Death Index. Land suggested that the amount of additional funding needed to pay for the current system is actually relatively modest—perhaps some $10 million annually, with $20 million in start-up costs for reengineering efforts in the states most in need. Land concurred with Schwartz that the VSCP would benefit from greater attention to demonstration of value and return on investment; the system needs to figure out how to market its products to a broader array of partners, so that NCHS is not alone in carrying the load at the federal level.

Nancy Krieger (Harvard School of Public Health) suggested that two concerns be kept in mind, regardless of which type of decision is pursued. One is that equity has to be at the core of any considerations; the second is that it is important to view the current situation in a broad historical perspective as well as in an international perspective.

Krieger said that reduction of health inequities is only possible if the nation's health data systems collect the data needed to understand the problem. Concerns about inequities in health along the lines of race, ethnicity, and socioeconomic position are central to the public good; hence, it is critical that changes to the vital statistics system not affect the ability to obtain those data. She noted that the idea of going to a sample-based approach for records collection would be a serious impairment to understanding inequities, undersampling or excluding some populations and oversampling others.

With regard to broad historical perspective, Krieger said that she was thinking about the debates that occurred in the 1930s when officials began to see the use of having not only vital statistics, but also more general data on health status. With that concern came the first national health survey, originally performed to assess the impact of the Great Depression. During that period, the nation came to realize that it needed different types of data for different purposes, and that different data streams can be mutually informative. For this reason, Krieger expressed concern about seeing NCHS data collections (including the vital statistics) pitted against each other in terms of funding needs or priorities.

The VSCP is sometimes criticized for having a somewhat 19th century approach with its emphasis on certificate processing, but Krieger suggested that there is much to be learned from the perspective that drove the development of statistical science in the 19th century. During that era, disciplines had not yet fractured to the extent they have now—there was no real distinction between statisticians, epidemiologists, and population demographers—so statistical work tended to bring together disparate aspects of government and research. Krieger argued that, going forward in vital statistics, there needs to be an emphasis on interdisciplinary research and intersectoral work; this amounts to returning to historical roots in the development of statistics writ large, as well as the emergence of the population health field.

Though not part of the final workshop session, a thread of discussion from an earlier workshop panel is particularly relevant to the notion of future discussions. At the end of the opening session on uses of vital statistics data, NCHS director Ed Sondik challenged the speakers to pick the one area for research or the one aspect of the VSCP that they felt was most in need of remedy, going forward. The speakers responded:

- Nancy Krieger (Harvard School of Public Health) commented that collaborative research involving academic researchers and staff of the relevant government agencies, together with staff in the state and local health departments, would be greatly beneficial for understanding key data issues. She particularly noted the status of population estimate methodology—understanding the denominators of vital statistics rates. She added a cautionary note that this kind of methodological research is not always fully appreciated, observing that her geocoding project work (see Chapter 2) was originally funded by the National Institute of Child Health and Human Development at the National Institutes of Health (NIH). When the program was up for renewal, NIH was less receptive to methodological work on public health surveillance. Though funding for such collaborative work may be difficult to secure, it could provide a better foundation for the system in the future.

Krieger also noted infrastructure developments currently being implemented in the Canadian province of Manitoba, along the lines of integrated systems being developed in Indiana. These systems come closer to integrating data systems across the life course, following people over time and permitting linkage of records as people go through different health systems. Work to figure out how to do the same thing more broadly in the United States would be a useful goal for infrastructure development, akin to Madans' list of priorities for future directions.

- Peter Van Dyck (U.S. Department of Health and Human Services) commented that the pressing concern is timeliness. For infant deaths, for example, the credibility of the system is undermined when the department is always reporting 3- and 4-year old data. Resources and efforts specifically geared to curbing data-release lag times should be the top priority, he said.

- Consistent with Krieger's broad historical and international perspective, Richard Rogers (University of Colorado) suggested that the system would benefit from serious attention to international comparisons, as well as state and local comparisons. He also noted that interfaces between the vital records reporting systems and systems for reporting deaths in special circumstances—military deaths and victims of natural calamities and other catastrophes—also merit exploration.

In summary comments on the workshop, Charles Rothwell (director of vital statistics, NCHS) commented that the VSCP faces great challenges, but expressed confidence that they can be overcome. He agreed with the general gist of Land's comments—that the nation gets a great deal from the current system that is not fully appreciated—but also agreed that there is still a set of major issues. Clearly, timeliness is a major issue for the system, both for the delivery of its current products and for the prospects of health surveillance or other new activities. He concurred that NCHS and the system are not doing enough to continually evaluate data quality, going back to check the accuracy and consistency of long-standing data items on the birth and death certificates, much less the new items in the 2003 revisions. Rothwell said that there has to be more to the vital statistics system than just funding the states and hoping they do well: Simply allocating more money to the VSCP is not necessarily the answer. He said that what needs to happen is to have programs in place that will help states that are weak in certain areas so that they can improve their systems; there must also be a commitment for VSCP finances to be appropriately and effectively used.

Workshop planning committee chair Louise Ryan (then, Harvard University Department of Biostatistics) closed the workshop by expressing the hope that this workshop and the discussions it spurred would be a catalyst for future action. The vital statistics system needs to be understood as a critical part of the nation's scientific infrastructure, and building awareness of the system's strengths and limitations is essential to continuing to provide vital data for national needs.

# Appendixes

– A –

# The U.S. Vital Statistics System: The Role of State and Local Health Departments

*Steven Schwartz*[*]

## INTRODUCTION

This paper describes the role of state and local health departments in the registration, processing, and analysis of vital events in the United States. It details the major responsibilities of offices of vital statistics and vital records and how vital events that are reported locally become state and national data. Many of the functions are basic to any data collection system, such as ensuring complete, accurate, and timely reporting. It is especially challenging in an environment that involves creating legal documents that prove U.S. citizenship, conducting public health surveillance, and protecting national security. This paper also explains the paper-based and automated systems currently in use and how they are evolving into web-based paperless systems that present opportunities for improved data quality and electronic verification and exchange of vital-event information.

---

[*]Steven Schwartz, Ph.D., registrar, Bureau of Vital Statistics, New York City Department of Health and Mental Hygiene, and past president, National Association for Public Health Statistics and Information Systems (NAPHSIS)

## BACKGROUND

In the United States, all vital events are local. What we see as national vital statistics start out as births, deaths, terminations of pregnancy, marriages, and divorces that are registered locally. There are 57 vital registration jurisdictions in the United States: The 50 states, 5 territories (Puerto Rico, U.S. Virgin Islands, Guam, American Samoa, and the Northern Mariana Islands), the District of Columbia, and New York City. Each of the 57 jurisdictions has a direct statistical reporting relationship with the National Center for Health Statistics (NCHS). Some of the states have centralized vital records offices, but most have local registrars who receive, register, and issue certified copies of vital records. There are over 6,000 local vital registrars nationwide.

Each of the jurisdictions is also a member of NAPHSIS, the National Association for Public Health Statistics and Information Systems. The association was formed in 1933 to provide a forum for the study, discussion, and solution of problems related to these programs in the respective members' health departments. NAPHSIS provides national leadership and advocacy on behalf of its members to ensure the quality, security, confidentiality, and utility of vital records and health statistics, as well as their integral role in health information systems, for monitoring and improving public health.

NAPHSIS represents the interests of jurisdictions on cross-cutting vital records and health statistics matters. These interests include national standards for data collection, exchange, issuance, and verification and electronic systems to collect, maintain, and disseminate records and data in a way that is timely, accurate, and protects confidentiality and security. NAPHSIS creates and fosters communication, best practices, training opportunities, and sharing of information across all jurisdictions. NAPHSIS works to build confidence in the integrity of public health and vital records systems and the data produced.

NAPHSIS also educates and negotiates on behalf of the jurisdictions with NCHS, the Social Security Administration (SSA) and other federal and national partners to ensure fair and equitable support and compensation for all jurisdictions, regardless of size or geographic location.

## MAJOR RESPONSIBILITIES OF VITAL STATISTICS OFFICES

Some jurisdictions began recording births, deaths, and marriages as early as 1632 (Hetzel, 1997). Later, this civil registration function expanded to include collection of public health statistics, beginning with death data. Jurisdiction vital records and statistics offices have long had responsibility for these two essential governmental functions: the civil registration of vital events and the collection of public health data. It is now recognized that vital records offices have a third, equally important function: helping to ensure national security. The 9/11 Commission Report recognized that birth certificate issuance can create risks and that it is necessary to devise standards for when a birth certificate can be issued (National Commission on Terrorist Attacks Upon the United States, 2004).

## Civil Registration

Jurisdiction and local vital records offices are enormous customer service operations. Over 11 million vital events are reported annually in the United States, including 4.3 million births, 2.4 million deaths, 26,000 fetal deaths (over 20 weeks gestation), 2.1 million marriages, and 1.1 million divorces (Sutton, 2008; MacDorman et al., 2007).[1] In addition, 1.2 million induced terminations of pregnancy are reported (Guttmacher Institute, 2008). Depending on jurisdiction laws, most of these events are reported directly to state and local vital statistics offices in the form of certificates of birth, death, marriage, and divorce and reports of spontaneous and induced termination of pregnancy.

State and local vital records offices, as the custodians of the records, must register and process these original, generally confidential records. Jurisdiction systems vary widely and by certificate type, which ranges from certificates that may be printed on 100 percent cotton paper with wet signatures for archiving to state-of-the art paperless records received through a secure Internet transmission with a biometric authentication. Regardless of the method, each record is unique, must be logged in, evaluated for data quality, and assigned file numbers.

The data on the records serve two purposes. For the legal registration of the record, it is essential to have accurate data to prepare certified copies and to create an index for retrieving the records. Vital records offices have the responsibility to maintain and produce certified records that are a true copy of the original. Data entry or retrieval errors would prevent vital records offices from discharging this core responsibility. For public health reporting and analysis, accurate demographic and medical data are equally important.

Because vital event certificates are legal documents, it is not trivial to make changes to them after they have been registered and filed. If families want to correct or amend a record, they must follow detailed procedures prescribed by the jurisdiction. Typical examples of corrections are spelling, date, and typographical errors. Documentation is generally required to prove what information is correct. Examples of birth record amendments include adoptions, adding a father's name, legal name changes, and changes of sex for transgender applicants. These amendments to birth records gener-

---

[1] The divorce figure is calculated using the rate of 3.6/1,000 population of 301.5 million, and excludes data for California, Georgia, Hawaii, Indiana, Louisiana, and Minnesota from the numerator and denominator.

ally require court orders or other legal instruments, such as an acknowledgment of paternity. All of these corrections and amendments must be treated carefully, to ensure that the documentation is proper and to prevent fraud.

A core function of vital records offices is issuing certified copies of birth and death certificates. This intrinsically local or state governmental function is both a drain on vital statistics offices' human resources and a source of revenue for jurisdictions. The revenue is substantial. Birth certificate fees among states in 2007 ranged from $9.00 (Florida) to $30.00 (New York), with an average price of about $15.00 per certified copy. Collectively, jurisdictions issue millions of certified copies of birth and death certificates annually. Unfortunately, the revenue often goes to a state's general revenue fund and not back to the program. Thus, personnel are often diverted from other work to meet important—and highly visible—customer service needs. As a result, less visible functions, such as quality assurance, training, and hospital site visits, may receive less attention or fewer resources.

## Public Health Data Reporting and Analysis

Every jurisdiction is responsible for the receipt, processing, quality, and analysis of its data. Vital-event data are derived from the certificates filed with each jurisdiction and become the vital statistics of that jurisdiction. They also become part of the nation's vital statistics through a cooperative agreement with NCHS, the Vital Statistics Cooperative Program (VSCP). It is cooperative because the collection of vital statistics is a function of the jurisdictions and not the federal government. Each jurisdiction has a contract with NCHS to provide data that meet the NCHS national standards for consistency, quality, and timeliness.

All jurisdictions have vital-event certificates that largely follow the U.S. standard certificates of live birth and death and the standard report of fetal death. There is no longer a U.S. standard report of induced termination of pregnancy (induced abortion), which was eliminated from VSCP funding in 1995. NCHS also stopped collecting detailed marriage and divorce data and now only publishes the monthly counts received from those states that report. The U.S. standard certificates and report are developed as a collaborative effort between the jurisdictions and NCHS. They are recommendations, not law. Each jurisdiction adapts the U.S. standards to meet its local needs. Some of these needs are for registration purposes, such as detailed funeral home or burial information on a death certificate. Other needs are to enhance public health data collection, such as adding a question on the birth certificate for foreign-born parents: If you were born outside the United States, how long have you lived in the United States?

Vital statistics data are tremendously valuable because the data represent the universe of events, not a sample. Vital-event reporting is mandatory, as it is with many reportable diseases and conditions. However, unlike other reportable events, reporting is virtually complete for births and deaths because families need certified copies for myriad legal, administrative, financial, and governmental purposes. Fetal death and induced termination reporting, although equally mandatory, are underreported because families do not have the same need for certificates.

Vital statistics systems are important surveillance systems and are becoming increasingly valuable as electronic birth and death registration systems help speed data acquisition and processing. Birth and death records are generally required by jurisdictions to be filed within 5 days of the event. This requirement, coupled with the needs of families to obtain certified copies of birth and death certificates quickly, enables jurisdictions to use their systems for surveillance. Electronic death registration systems may prove especially valuable for disease surveillance in the event of an influenza pandemic or a terrorist attack.

One of the great benefits of the jurisdiction-based system is that the jurisdictions know their data and their data providers best. At the national level, NCHS must rely on jurisdictions to know their data, monitor quality, and work with their data providers. Vital records and statistics staff see the records as they come in, either electronically or on paper. This oversight presents opportunities to evaluate data quality and to look for systematic errors caused by specific hospitals, birthing centers, medical examiners and coroners, and funeral directors. Jurisdiction staff also interact directly with families who are applying for certified copies and for corrections and amendments to records. Complaints and errors recognized here may point to filing and procedural problems in the providing institutions.

Unfortunately, data providers do not work for vital statistics. They are the physicians, nurse midwives, medical examiners, coroners, funeral directors, hospital and clinic directors, clerks, and temporary staff who complete the worksheets and certificates that eventually become data. The challenge for vital statistics staff is to maintain and improve data quality. That work includes training, educating, and querying providers about birth and death data and causes of death, and conducting active surveillance for fetal deaths and induced terminations of pregnancy.

Underreporting of deaths among extremely-low-birth-weight infants is a special concern because of its impact on infant mortality rates (Paulson et al., 2007). Underreporting of these infant deaths is likely because of extremely short lives and confusion over whether to report them as live-born infants or fetal deaths. It is the responsibility of jurisdiction vital statistics offices and staff to work with and know their data so that they can conduct rigorous follow-back to data providers on very-low-birth-weight infants.

## National Security

Birth certificates are breeder documents. In the United States, birth confers citizenship, and birth certificates constitute the proof. Thus, birth certificates are used by SSA to generate Social Security numbers, by the U.S. Department of State as evidence for passports, and by state departments of motor vehicles to issue driver's licenses.

Vital records offices always have had to protect against fraud. Alterations to birth certificates can be used to change identities or to steal them. Death certificates may be altered to commit fraud against insurance companies or to escape arrest warrants.

The events of 9/11 and escalating identity theft have sensitized all vital records offices to the risks associated with birth certificate fraud. However, the vital records system was not designed to protect national security, and it is currently inadequate for the task. For example, to protect against identity theft, it is important to match death with birth certificates and then mark the birth certificate "deceased." To be effective against fraud, this matching must be done quickly. This is not possible with current systems, since all births and deaths are neither reported nor processed fast enough to enable rapid matching. Furthermore, since a person often dies in a jurisdiction different from that of birth, a national system would be needed to permit rapid data sharing and matching. Today, most of that sharing is not automated and often done more than a year after a death.

## CURRENT AND FUTURE SYSTEMS

Vital records offices are evolving from their original paper-based systems to web-based paperless systems. Vital records certificates for the public still will be printed on paper with a "raised seal" to indicate authenticity for the foreseeable future, but most of the systems that lead up to the paper are changing. NAPHSIS, with support from NCHS and SSA, worked with the jurisdictions to develop national models and standards. The national standards facilitate jurisdictions in building these complex automated systems, thereby promoting uniformity and saving resources and time.

Paper certified copies will be subject to electronic verification through the Electronic Verification of Vital Events (EVVE) system to confirm that they match true records on file in a vital records office. The State and Territorial Exchange of Vital Events (STEVE) system will greatly facilitate exchange of certificate information among jurisdiction vital records offices, NCHS, and other partners. Electronic birth and death registration systems will improve timeliness, efficiency, and data quality and will reduce the need for corrections. Electronic death registration systems present opportunities for improving cause-of-death reporting, although there have not yet been breakthroughs. Currently, electronic systems have internal consistency checks, but they still rely on careful and accurate collection and entry of data.

### State and Territorial Exchange of Vital Events (STEVE)

Generally, jurisdictions send all birth, death, fetal death, and induced abortion records that occurred to nonresidents to the jurisdiction of usual residence. Death records are also sent to the jurisdiction of the decedent's birth and, for decedents less than 1 year of age, birth records are sent on request to the jurisdiction where the death occurred.

Most jurisdictions and several Canadian provinces have signed an agreement that specifies the terms and conditions under which vital events data may be exchanged and used. However, the enforcement of its terms and conditions is not automated and is completely dependent on the vigilance of individual staff. Other formal agreements govern the exchange of data between jurisdictions and NCHS, SSA, other agencies of the Centers for Disease Control and Prevention, and other internal and external organizations. Some or all of these organizations will become trading partners and users of the STEVE system.

NAPHSIS has developed standard record layouts for electronic exchange. However, most jurisdictions currently do not exchange records electronically because of lack of resources and an efficient exchange platform. Instead, they may send copies, computer abstracts, or line listings to each other. Forthcoming Intelligence Reform and Terrorism Prevention Act regulations may require jurisdictions to match birth and death records beyond current practices, which will affect interjurisdictional exchange procedures. Additionally, the threat of natural or man-made disasters indicates the need to report deaths nationally on a timelier basis. Therefore, the need for an improved electronic exchange platform is both important and immediate.

STEVE is a Public Health Information Network–compliant, secure messaging system currently under development by NAPHSIS that will allow jurisdictions to electronically trade the vital-event information they are currently sharing. STEVE will be installed at each participating jurisdiction within its secure firewall. Each jurisdiction will be allowed to configure the data exchange and use rules to meet its own regulatory requirements and business practices. The detailed rules governing the exchange and use of the data will be embedded in a specialized, configurable software application. Jurisdictions may also use STEVE to send data to other approved trading partners, such as NCHS, and to authorized public health agencies and programs such as immunization, newborn screening, and birth defects.

Receiving jurisdictions will be able to configure multiple mailboxes on STEVE for internal state agencies and programs, such as newborn hearing screening, birth defects, cancer registries, child support enforcement, voter registration, and other programs with which they have an official data-sharing relationship. These mailboxes may be used to distribute internal records as well as interjurisdictional records, thereby eliminating the need to extract and distribute multiple data sets for programmatic use.

Development began in February 2008 and a start-up trading-partner community is expected to go live by December 2008. Expansion of the trading-partner community will take place over a 2- to 3-year period, and is expected to include all 57 vital records jurisdictions, NCHS, and selected additional partners.

### Electronic Verification of Vital Events (EVVE)

Many federal and state agencies rely on birth certificates for proof of age, proof of citizenship and identification for employment purposes, to issue benefits or other documents (e.g., driver's licenses, Social Security cards, and passports), and to assist in determining eligibility for public programs or benefits.

NAPHSIS has developed and implemented an electronic system that allows immediate confirmation of the information on a birth certificate presented by an applicant to a government office anywhere in the nation irrespective of the place or date of issuance. Authorized federal and state agency users via a single interface can generate an electronic query to any participating vital records jurisdiction throughout the country to verify the contents of a paper birth certificate or to request an electronic certification (in lieu of the paper birth certificate). An electronic response from the participating vital records jurisdiction either verifies or denies the match with official state records. It also flags positive responses when the person matched is now deceased. As designed, queries could be generated and matched against 250 million birth records in vital record databases nationwide.

EVVE has been tested in pilot projects with state motor vehicle administration offices and federal Medicaid offices, the Office of Personnel Management, and SSA. The following states are online with EVVE: Arkansas, Hawaii, Iowa, Kentucky, Minnesota, Mississippi, Missouri, Montana, North Dakota, South Dakota, and Utah.

## CONCLUSIONS

The U.S. vital statistics system is a cooperative effort of the 57 vital registration jurisdictions, NAPHSIS, and NCHS. It relies heavily on the work of each of the jurisdictions to meet its local vital registration and public health data needs. It also relies on the jurisdictions to meet national standards for data quality and timeliness as defined in the VSCP contracts. Inadequate jurisdictional resources and local demands for customer service compete for the limited staff necessary to meet national standards. New electronic systems for birth and death registration and for data sharing and verification will help, but they also will be competing with new national security requirements to protect birth certificates. In addition to new technology, leadership at the local, state, and national level is acutely needed to build a stronger national system.

# – B –

# The U.S. Vital Statistics System: A National Perspective

*National Center for Health Statistics*

### INTRODUCTION

The development and maintenance of a system to produce national vital statistics based on the local registration of vital events was a major accomplishment of the United States during the 20th century. The National Center for Health Statistics (NCHS) is the federal agency legislatively mandated to produce national health statistics based on this cooperative, decentralized system in which data from more than 6 million vital-event records are collected each year by all states and U.S. territories and transmitted to NCHS for processing and dissemination. Looking to the 21st century, the local, state, and federal government organizations that comprise the National Vital Statistics System (NVSS) are engaged in significant changes through redesign and automation that should dramatically improve the performance and security of the system. From data collection to data processing and dissemination, efficiencies are being sought in an effort to improve both timeliness and quality at all levels of the system and to reduce costs.

# THE NATIONAL VITAL STATISTICS SYSTEM

In the United States, legal authority for the registration of births, deaths, marriages, divorces, fetal deaths, and induced terminations of pregnancy (abortions) resides individually with the states (as well as cities in the case of New York City and Washington, D.C.) and Puerto Rico, the Virgin Islands, Guam, American Samoa, and the Commonwealth of the Northern Mariana Islands. In effect, these 57 jurisdictions are the full legal proprietors of the records and the information contained therein and are responsible for maintaining registries according to jurisdiction law, including issuing copies of birth, marriage, divorce, and death certificates.[1]

As a result of this state[2] authority, the collection of registration-based vital statistics at the national level has always depended on a cooperative relationship between the states and the federal government. Since its inception in 1960, NCHS has been the organization responsible for the federal aspects of this enterprise. NCHS has legislative authority and is mandated under 42 U.S.C. § 242k, Section 306(h) of the Public Health Service Act to collect vital statistics annually: "There shall be an annual collection of data from the records of births, deaths, marriages, and divorces in registration areas. The data shall be obtained only from and restricted to such records of the States and municipalities which the Secretary, in his discretion, determines possess records affording satisfactory data in necessary detail and form." Currently this data collection is limited to data from birth and death records (including fetal deaths), as NCHS discontinued the collection of individual-record marriage and divorce reports after 1995.

The states are collectively represented in their dealings with the federal government by the National Association for Public Health Statistics and Information Systems (NAPHSIS). NAPHSIS is a professional organization whose members include primarily, but not exclusively, the vital registration and statistics executives and other employees of state registration offices. In addition to providing the states with a common point of contact with the

---

[1] The NVSS is based on the local registration of vital events. For births and deaths, this typically works, in outline form, as follows: Demographic information on the birth certificate is provided by the mother at the time of birth, and medical and health information is based on medical (i.e., prenatal care, hospital, etc.) records. Demographic information on the death certificate is provided by the funeral director based on information supplied by the informant (usually the next of kin). A physician, medical examiner, or coroner provides medical information on cause of death. The completed birth and death certificates are registered with the local or state registrar by, respectively, the hospital records officer or the funeral director. The local registrar subsequently files the records with the state vital registration office, which codes and keys the data and transmits a copy of the electronic file to NCHS. The state offices are responsible for maintaining archival copies of records and for issuing certificate copies. Upon receipt at NCHS, the data are edited and assembled into national files for analysis and publication. NCHS sets uniform standards for data that will be collected and for item coding.

[2] In the following text, the word "state" will be used to refer to all jurisdictions.

federal government and numerous other professional organizations, NAPHSIS also facilitates interstate exchange of ideas, methods, and technology for the registration of vital events and dissemination of vital and other public health statistics. NAPHSIS's progenitors date back to 1933, when it was organized as the American Association of State Registration Executives; among other name changes, it was renamed the Association for Vital Records and Health Statistics (AVRHS) in 1980.

## MILESTONES IN NATIONAL VITAL STATISTICS

The civil registration of births, marriages, and deaths has a long history in the United States, beginning with a registration law enacted by Virginia in 1632 and a modification of this law enacted by Massachusetts in 1639. The original impetus for these laws was the protection of individual rights, particularly relating to ownership and distribution of property, and not for statistical uses. However, with the rise of industrialism and urbanization in the 19th century, and the associated epidemics of diseases (typhus, yellow fever, cholera) arising from poor sanitation and polluted water, information from mortality records was used to develop support for sanitary reform and public health in general. The names of Edwin Chadwick (1800–1890) and Dr. William Farr (1807–1883) stand out in this effort in England, and that of Lemuel Shattuck (1793–1859) is prominent in the United States. Shattuck succeeded in steering through the Massachusetts legislature in 1844 a bill that required central state filing, provided for standard forms, fees, and penalties; specified types of information including causes of death; and lodged responsibility for each kind of record in designated officials (for more details, see Hetzel, 1997).

In the late 1840s, the newly formed American Medical Association (AMA) began to promote the use of mortality statistics in the study of health conditions of the U.S. population. In 1847 the AMA addressed memorials to state legislatures on the need for registration laws. To obtain national data, the decennial censuses in the latter half of the 19th century included questions about vital events, but the method was recognized as inefficient and the results as deficient. Accordingly, in 1902, when the U.S. Bureau of the Census was made a permanent agency of the federal government, the legislation authorized the director of the Bureau to obtain, annually, copies of records filed in the vital statistics offices of those states and cities having adequate death registration systems and to publish data from these records. A few years earlier, the Bureau had issued a recommended death reporting form (the first "U.S. Standard Certificate of Death") and requested each independent registration area to adopt it as of January 1, 1900. Those areas that adopted the form and whose death registration was 90 percent complete were to be included in a national death-registration area that had been established in 1880. In 1915 the national birth-registration area was established, and by 1933 all states were registering live births and deaths with acceptable event coverage and providing the required data to the Bureau for the production of national birth and death statistics.

Fetal deaths of 20 weeks of gestation and greater have been a reportable component of U.S. vital statistics since the 1920s. Following the Supreme Court's 1973 ruling on *Roe v. Wade*, the need for a separate reportable component on abortion was recognized and in 1978 the first standard report of induced termination of pregnancy (ITOP) was introduced. NCHS discontinued the collection of ITOP data from the states in 1994 as a cost saving measure. Federally compiled data on abortions are available from the National Center for Chronic Disease Prevention and Health Promotion's Division of Reproduction Health which has had a voluntary arrangement with states since 1969 to collect tabulated data on abortions, including the characteristics of abortion patients, occurring in the states. The current Model Law[3] definition of fetal death specifically excludes ITOPs. There are more than 25,000 fetal deaths in the U.S. each year, similar to the number of infant deaths, and fetal death reporting remains an integral component of NVSS at NCHS.

Registration areas for marriage (MRA) and divorce (DRA) statistics were established in 1957 and 1958, respectively. The MRA began with 30 states (excluding New York City) plus Alaska, Hawaii, Puerto Rico, and the Virgin Islands; when detailed marriage data were dropped in 1996 there was a total of 45 registration areas. The DRA began with 16 states, the District of Columbia and the Virgin Islands; when detailed divorce data were dropped in 1995 there was a total of 33 registration areas. Beginning in 1996, NCHS has regularly published monthly counts of marriages and divorces in the reporting jurisdictions. Currently, all states are reporting counts of marriages to NCHS. Forty-four states and the District of Columbia are reporting counts of divorces (counts are not reported by California, Georgia, Hawaii, Indiana, Louisiana, and Minnesota).

---

[3] A prime example of cooperative developmental work is the Model State Vital Statistics Act and Regulations, 1997 Revision; see http://www.cdc.gov/nchs/data/misc/mvsact77acc.pdf. NCHS facilitated the process of revising the Model Act and Regulations by convening a Revision Working Group and bringing in various experts that had an interest in the registration system. This is the fifth edition of the Model Act (the first was produced in 1907) and the third edition of the Model Regulations (the first was in 1973). The Model Act and Regulations provide a legal guide to states that are considering refining their laws. The report on the 1997 revision provides a useful reference on terminology, recommended registration practices, disclosure and issuance procedures, and other functional aspects of a state registration system. The latest revision of the Model Act and Regulations contains provisions that allow states, which implement the relevant sections of the Model, to easily incorporate technological advances in records and information management.

In 1946, responsibility for collecting and publishing vital statistics at the federal level was transferred from the U.S. Bureau of the Census to the Federal Security Administration, and later (1953) to the National Office of Vital Statistics (NOVS) in the U.S. Public Health Service. In 1960 NOVS was merged with the National Health Survey to establish NCHS. The Division of Vital Statistics (DVS) was created in a 1963 reorganization of NCHS. In 1987, NCHS became part of the Centers for Disease Control and Prevention (CDC) in the U.S. Department of Health and Human Services (HHS).

## THE VITAL STATISTICS COOPERATIVE PROGRAM

Prior to 1971, the federal government reimbursed the states for use of their vital records for national statistics at the rate of 4 cents per record. The records were produced by local officials and transmitted to state vital registration offices for permanent maintenance in central state files and production of state and local vital statistics. Basic standards for definitions, data quality, and methods of collecting, processing, and analyzing vital statistics were in place. The states forwarded microfilm copies of the records to NCHS, which edited, coded, tabulated, and published vital statistics for the nation. This process began to change in 1971, when the first state, Florida, transmitted magnetic tapes of state-coded data to NCHS, coded according to NCHS specifications. By 1973, six states had entered into contracts with NCHS to provide computer tapes of birth and demographic death data under a new formal arrangement called the Vital Statistics Cooperative Program (VSCP).

During this early development period, priority was given to birth and demographic death data, and states were brought into the VSCP when they were technically ready and sufficient funds were available in NCHS to establish a contract. Some states provided tapes to NCHS prior to availability of funding. Priority for contracts was given to non-VSCP states (rather than to adding subcomponents to states already under contract).

The number of states submitting to NCHS one or more components of data under the VSCP gradually expanded as funding permitted. By 1985 all states, the District of Columbia, New York City, Puerto Rico, and the Virgin Islands were submitting birth data and demographic death (not cause of death) data on tape. In 1995, 42 states and the District of Columbia were submitting medical death data on tape. Except for periodic problems in reporting, all jurisdictions except West Virginia, Virgin Islands, American Samoa, Guam, and the Northern Marianas are currently submitting medical death data in the format of the NCHS Mortality Medical Data System (described below).

The original VSCP contracts were negotiated individually with each state, but as states entered the VSCP, it soon became apparent that a rationale had to be developed to simplify the process and assure funding equity among the states. In 1981 the director of NCHS established for this purpose a work group comprised of selected staff from NCHS and state representatives appointed by the president of the AVRHS. The work group's assigned tasks were to (1) define the state/local activities involved in producing vital statistics for all levels of government, (2) figure the cost of those activities, (3) establish a rationale for determining the federal share of that cost, and (4) develop a standardized funding formula for the VSCP contracts. The end result of this effort was "a cost formula limited to the accepted level of effort necessary to carry out the in-scope functions of the vital statistics contracts." This formula was used to distribute available funds to all states under VSCP contracts until 1987.

A second VSCP work group was established in 1986 to review the experience with the existing rationale and cost formula and develop recommendations for revisions. This work group recommended (1) updating and simplifying the cost formula to focus on the collection of standard data sets rather than an item-specific approach, (2) eliminating reference to the "federal share" in favor of a funding level derived from a base level with annual cost-of-living adjustments, (3) providing for funding additions to cover the cost of collecting and processing new items of data, (4) providing for reduction in scope of contract to offset reductions in federal funding, and (5) requiring a state to report all minimum basic data-set items to receive full contract funding.

The third joint NCHS/AVRHS work group to review the VSCP cost formula was established in 1992. This work group made few changes in the previously established VSCP provisions, recommending principally that the cost formula for the years 1995–1999 use the staffing and salaries data collected from the states in 1995 with overall application of annual cost-of-living adjustments. This work group had extensive discussions of ways to improve timeliness of data production, ultimately recommending that states should send data to NCHS as soon as records were received and initially processed rather than waiting for full quality control to be completed; updated records were to be transmitted as amendments were processed.

An important outcome of the emphasis on timeliness by the 1992 work groups was NCHS's introduction of a new statistical series, based on a new approach to collecting and processing vital statistics data. Beginning with 1995 data, NCHS instituted an annual publication of preliminary vital statistics data based on a very substantial sample (80–90 percent) of records, including detailed tabulations from the natality as well as mortality files. Consequently, in January 1998, NCHS ceased publication of provisional mortality data based on the Current Mortality Sample (CMS), a sample of

10 percent of death records received each month from the states and coded and classified by federal staff. The Census Bureau had instituted the CMS in 1943 in response to concerns about the threat of epidemics and the possibility of a general decline in national health resulting from wartime living conditions.

In the year 2000, a new 5-year VSCP contract was established and simplified the cost formula that had been in effect since 1995, and this contract was extended in the years since 2006. For 2007 and 2008, CDC changed the procurement mechanism to purchase orders, but retained the funding distribution determined by the cost formula of 1995. Thus, the funding distribution to states remains basically the same as in 1995, with the total funding for the VSCP increasing each year by a general cost-of-living factor. Over the last 13 years, many changes in systems and procedures have taken place at the data provider and state levels, and it is unknown whether the current payment to states through the VSCP reflects an increase or decrease in the NCHS share of the current state cost of data collection and processing.

The recommendations of the 1986 VSCP work group were applied in the mid-1990s as a result of funding reductions for NCHS. To adjust total contract funding downward, NCHS eliminated several VSCP components, ceasing collection of all data on induced terminations of pregnancy as of 1995 and detailed data on marriages and divorces in 1996. Several data items were also eliminated from the minimum basic data sets for natality (1-minute Apgar score, date of last live birth and last fetal death, and education of father). For the mortality minimum data set, the autopsy item was removed, although later restored in 2002. As of 2008, however, other eliminated items and data sets have not been reinstated.

Recent level-funding budgets for NCHS and other organizations in the federal government have created another crisis, but so far no data elements or components have been eliminated. Instead, the contracts (or purchase orders) have been funded to purchase data for shorter and shorter periods of time. Using the full VSCP cost for each year determined by successive cost-of-living adjustments, a daily "burn rate" is calculated for the fiscal year, and the contracts are then set up to cover the number of days in the year that the burn rate will purchase with the actual funding available.

Complicating this picture is the issue of funding needed to cover the cost of important new medical data items in the 2003 revisions of the U.S. Standard Certificates of Birth and Fetal Death. NCHS has held negotiations with NAPHSIS on this issue but, given NCHS's inability to fund even a full year of the current minimum basic data set, there is little likelihood that funds will be available for additional data items in the upcoming budget years. This is a particularly unfortunate situation because the new medical and health data items are especially important for current issues in reproductive health. Currently, NCHS and NAPHSIS are engaged in the early stages of discussions to develop strategies for disseminating data on these items, with focus on assessing data quality; the hope is that these collaborations will generate interest and support for steady and dependable funding.

## COOPERATIVE ACTIVITIES IN SYSTEM DEVELOPMENT

### E-Vital Initiative

E-Vital was among the first of the 24 presidential e-government initiatives promulgated by the George W. Bush administration and monitored continuously by the U.S. Office of Management and Budget (OMB). The purpose of E-Vital was initially to establish common electronic processes for federal and state agencies to collect, process, analyze, verify, and share death and birth record information, thereby reducing the burden on state agencies for reporting vital events and increasing the quality of the vital-event information being recorded. Until fiscal year 2008, the Social Security Administration (SSA) had the lead for this initiative and HHS, among other federal agencies, was a partner.

During the initial phase of the initiative, a pilot system was developed to demonstrate the practicality of Electronic Verification of Vital Events (EVVE) by creating an online system that would enable federal agencies to query state vital records offices to verify the availability and accuracy of birth and death records, for example, to determine qualification for new or continuing entitlements. The EVVE system was developed and shown to work; however, determining the appropriate payment per transaction for states remains elusive.

### Electronic Death Registration (EDR)

The second portion of the E-Vital initiative, more important for vital statistics, was to support states in their efforts to reengineer their death registration process, i.e., implement EDR. Timely reporting of death information is critical for detecting and defining pandemic and other calamitous events. Yet, timely death reporting has been a major challenge for the mortality vital statistics system, primarily because state death registration systems have been essentially paper-based systems. Using grant funding (now exceeding $10 million) from SSA, states with assistance from NAPHSIS and NCHS have begun to implement EDR systems to improve the timeliness of fact-of-death information.

Since 2002, 40 of the 57 registration jurisdictions (50 states, New York City, the District of Columbia, and 5 U.S. territories) have either implemented an EDR or are in the process of implementing one. The SSA experience has successfully demonstrated that EDRs can provide fact-of-death information from the local level through the state to the federal level in a timely fashion. However, state reporting of cause of death through these systems continues to be a challenge. The critical needs now are twofold: (1) expanding EDRs within and to all states, and (2) improving the timeliness of cause-of-death reporting through EDRs.

## Intelligence Reform and Terrorism Prevention Act

In December 2004, the president signed into law the Intelligence Reform and Terrorism Prevention Act (IRTPA), which has the potential to boost efforts to implement electronic birth registration (EBR) systems as well as the EDR. Section 7211 of this act requires that the secretary of HHS establish minimum standards to improve the security of birth certificates. This section emanates directly from the 9/11 Commission's report, which provided a variety of recommendations on terrorism prevention including the appropriate use and method of obtaining copies of birth certificates (National Commission on Terrorist Attacks Upon the United States, 2004). Congress acted on this report with the passage of IRTPA, including Section 7211 which is the first federal statute to regulate vital registration practices of the states.

IRTPA specifically mandates three categories of minimum standards for vital registration, including standards on (1) the certification of birth certificates and the use of safety paper, (2) proof and verification of identity as a condition of issuance of a birth certificate, and (3) processing of birth certificate applications to prevent fraud. It also authorizes a grant program to assist states in meeting the federal standards and in computerizing their registration systems for the timely matching of birth and death records and noting the fact of death on the decedent's birth certificate. While funding was authorized for the states to implement the regulations, as yet no funds have been appropriated. It is anticipated that many states will need to modify some portion of their vital registration statutes in order to meet the proposed federal regulations.

HHS was given the lead in the vital registration portion of IRTPA and the secretary of HHS delegated that lead to CDC; DVS within NCHS is handling the effort. All federal agencies affected by the legislation, including the U.S. Justice Department, the U.S. Department of Homeland Security, the U.S Department of State, the U.S. Department of Transportation, SSA, the Government Printing Office, and HHS, along with state vital registrars, were brought together to discuss the regulations. As a result of a series of meetings in the summer of 2005, this group has provided DVS with recommendations for these regulations. Using these recommendations and the services of a contract legal team paid for by SSA, DVS staff drafted a set of proposed regulations and submitted them to HHS for its review and approval.

The primary thrust of the draft regulations will be to standardize security practices for states in the registration and issuance of birth certificates. The goal is to have a more secure and responsive vital registration process in every state by enhancing their system's infrastructure and providing secure electronic transmission of data within and between states and with federal agencies. The data will continue to be "owned" by the states; states will continue to manage and be responsible for vital registration; and there will be no federal database derived uniquely from this legislation.

Despite the primary thrust on security, CDC/NCHS has a policy interest in these regulations and thus specific reasons to work in concert with other agencies to develop the regulations and adjudicate differences. Aside from the broader interest served by improving security, the primary policy interest of CDC/NCHS is to advance a long-standing public health interest in more rapid statistical information that is collected through the registration of births and deaths. Vital statistics have been built on the vital registration process in each state, and the modernization of the process and infrastructure of vital statistics reporting (most specifically automating reporting at the source) can provide an early warning system in every community to track in real-time high-risk births and deaths of public health interest.

## COOPERATIVE ACTIVITIES IN DATA DEVELOPMENT

### U.S. Standard Certificates and Reports

Periodic revision of the U.S. Standard Certificates and Reports is a significant area of cooperation for NCHS and the state vital statistics offices, occurring generally every 10 to 15 years. This is a particularly important activity because it brings together various experts—data users, researchers, and policy makers, both public and private—to develop recommendations on the content of the certificates and reports that will be used in the registration of live births, deaths, marriages, divorces, fetal deaths, and induced terminations of pregnancy during the next decade. Although the states do not all adopt the U.S. Standard Certificates and Reports exactly as they are promulgated, the documents are generally employed with only minor changes and thereby succeed in promoting a high degree of uniformity and comparability among the states. In addition, VSCP contracts between NCHS and the states require collection of certain basic data items from the various certificates and reports.

The standard certificates have been the principal means for achieving the uniformity in information on which national vital statistics are based. The U.S. Bureau of the Census developed the first standard certificates for the registration of vital events—births and deaths—in 1900. To date there have been 11 revisions of the Standard Certificate of Live Birth, 10 revisions of the Standard Certificate of Death, 7 revisions of the Standard Report of Fetal Death (formerly Stillbirth), 3 revisions of the Standard Certificates of Marriage and of Divorce or Annulment, and 1 revision of the Standard Report of Induced Termination of Pregnancy. A published report (Tolson et al., 1991) describes the procedures followed in developing the revisions and the principal additions, modifications, and deletions of items for the 1989 revisions. This report also provides a history of the content of all certificates and reports since 1900.

## The 2003 Revisions

The latest certificate revision process began in January 1998 and the report of the evaluation panel was issued in April 2000. Addenda to the report, explaining changes made after the initial recommendations of the panel, were issued in November 2001 (Division of Vital Statistics, 2000). The revision process evaluated only the live birth and death certificates and the fetal death report. The revised certificates and report were originally scheduled for implementation in 2002. However, in consultation with NAPHSIS, NCHS decided to delay implementation until January 1, 2003, because of the complexity of changing automated systems in the states and the need to test the recommended changes before implementation. It was agreed that the fundamental goal should be to move from "a system primarily based on the flow of paper to the faster electronic registration of vital events. The Panel looked beyond designing new paper documents and concentrated on cultivating an appropriate vital statistics data base grounded in the electronic transfer of information" (Division of Vital Statistics, 2000). Moreover, it was recognized that the EBR systems in existence at the time were based on outmoded software and hardware and should be reengineered before new certificates were implemented.

## 2003 Changes in Collecting Data on Births and Fetal Deaths

With the 2003 revision of the birth certificate and fetal death report, the panel recommended the development of worksheets to collect demographic data from the mother and separately to collect medical and health information from the prenatal care records and the birth facility. This was a significant shift from previous data collection procedures that typically depended on obtaining all relevant information from the mother. But this recommendation was subject to demonstration that data could be obtained accurately from the prenatal care records and the birth facility records by hospital staff. A study was successfully conducted in collaboration with NAPHSIS that found the information could be effectively collected this way. DVS also sponsored the development of a "Guide to Completing the Facility Worksheets" for the Certificate of Live Birth and Report of Fetal Death, designed and developed to assist medical records personnel in completing the facility worksheets for births and fetal deaths (Division of Vital Statistics, 2006). All of this represented fundamental change in the ways vital statistics data had been collected in the past and focused direct attention on improving data quality.

States have not been able to uniformly implement the 2003 birth certificate revision as planned in January 2003. Instead, the states are switching to the revised birth certificate as they are able to obtain sufficient funds to reengineer their EBRs and to train hospital staff in abstracting data for the worksheets. Two registration areas adopted the 2003 revision in 2003, 7 additional areas adopted in 2004, 5 in 2005, 6 in 2006, for a total of 25 by the end of 2007 and 32 states as of 2008. Implementation of the revised Standard Report of Fetal Death is similarly being phased in, with a total of 16 registration areas adopted by 2007. NCHS must therefore maintain records for two different kinds of data sets between which some of the variables are not comparable. Further complicating the situation is that some states have implemented certificate and report revisions in the middle of a data year. In these cases, the states do not have a consistent statistical data set for the affected year. This creates an extraordinarily difficult problem for compilation of data files and for the analysis and dissemination of national birth and fetal death data. This will continue to be the case until all jurisdictions have switched to the newest certificate formats.

### 2003 Changes in Collecting Information on Deaths

A subgroup to evaluate the U.S. Standard Certificate of Death made several semantic changes to the standard death certificate and reorganized portions of the certificate, as appropriate, to ease the use of this document. Many of the subgroup's recommendations included wording changes or the addition of check boxes to existing certificate items to obtain more detailed information. In addition, the subgroup added items to the certificate to address public health concerns and issues associated with International Statistical Classification of Diseases and Related Health Problems, Tenth Revision (ICD-10), coding. Among other items, the subgroup added a question to collect information on whether tobacco use contributed to death, a question to collect information on the pregnancy status of female decedents, and a question to collect additional information on traffic deaths. As of 2008, 33 registration areas had adopted the new death certificate; most of these adoptions were paper versions because EDR systems generally do not yet cover any full jurisdiction for cause-of-death reporting.

## 2003 Changes in Collecting Information on Race and Ethnicity

As of January 1, 2003, federal programs were required by OMB to adopt revised standards for collecting and reporting racial and ethnic status. These standards were published in the *Federal Register* on October 30, 1997, as "Revisions to the Standards for the Classification of Federal Data on Race and Ethnicity."[4] The U.S. Census Bureau was one of the first federal agencies to implement the revised standards, incorporating in the 2000 decennial census a format for the race question that included 15 check-box items and 3 write-in lines, plus the instruction to "Mark one or more races to indicate what this person considers himself/herself to be." Subsequently, the Panel to Evaluate the U.S. Standard Certificates recommended that the revised 2003 standard certificates should have race and Hispanic origin questions nearly identical to those in the 2000 census in order to maintain comparability of the data collected in census and vital statistics. To facilitate coding and processing of multiple-race/Hispanic-origin data in a uniform manner for all vital statistics jurisdictions, NCHS has developed a computer system to code (with minimal manual intervention) and edit reported data. The system also bridges multiple-race data into the single race categories of the 1977 OMB race standard using a bridging algorithm developed by NCHS. For presenting vital statistics data by race in NCHS publications, all national tabulations use bridged race in place of multiple race and bridging will continue as long as some jurisdictions continue to use the old race standard. Moreover, NCHS will need to continue to use bridged-race population estimates for denominators to calculate rates.

## The Linked Birth/Infant Death Data Sets

The Linked Birth and Infant Death File (LBIDF) project is a major area of cooperation between NCHS and the state vital statistics offices. For analytical purposes, it is especially useful to combine information from the birth and death certificates for any infant that dies; the additional variables from the birth certificate make a much richer infant mortality database. In addi-

---

[4]The notice is posted on the OMB website at http://www.whitehouse.gov/omb/fedreg/ombdir15.html. The revised standard certificates, with the revised race and Hispanic origin formats, may be found by going to the following website, where the data collection, transmission, edit, and file layout specifications are also posted: http://www.cdc.gov/nchs/vital_certs_rev.htm. The NCHS code lists for race and origin are accessible at the following website: http://www.cdc.gov/nchs/data/dvs/RaceCodeList.pdf. A description of bridged race data from the 2000 census for counties, states, and nation is accessible at the following NCHS website: http://www.cdc.gov/nchs/about/major/dvs/popbridge/popbridge.htm. This report includes a report describing the bridging algorithm, including its development and characteristics. The goal is to eventually make the coding and editing algorithm available interactively on the web, so states can submit race and Hispanic origin data to the program and receive back the edited results without manual processing by NCHS.

tion, with an LBIDF it is possible to use race of mother for both numerator and denominator in an infant mortality rate thereby improving the reporting of race for infant mortality data. Because an infant born in one state may die in another, the child's birth and death certificates may be registered in different states. NCHS facilitates an interstate agreement to exchange infant death and birth certificates. LBIDF is produced in two different formats: birth cohort data and period data. The birth cohort data contain information for infants born in a particular year who died before their first birthday, either in that year or the following year. Beginning with 1995 data, NCHS also began producing period-linked file data. The period data include all infant deaths in a particular calendar year, linked to their respective birth certificates, whether the birth occurred in the current or the previous data year. The period format allows NCHS to release linked file data in a more timely fashion, since it is no longer necessary to wait for an additional year of mortality data to see if an infant died in the following data year. Thus, the period-linked file is a more effective tool for surveillance, while the birth-cohort-linked file is more suited to in-depth research projects. Birth cohort-linked file data are currently available for the 1983–1991 and 1995–2003 cohorts. Period-linked file data are currently available for 1995–2004 data years.

### Vital Statistics Follow-back Surveys

The vital statistics follow-back surveys conducted by NCHS depend entirely on the cooperation of state vital statistics offices. A number of such surveys have been conducted since the mid-1950s by NCHS and its predecessor, the National Office of Vital Statistics, usually in collaboration with other federal government agencies. The state offices support the surveys by obtaining approval from their own health departments and institutional review boards, as necessary, and by authorizing the use of copies of vital records in the sample. These surveys typically are based on a sample of vital records in an annual birth or death file. Questionnaires are sent to sources of information identified on the records, for example, to the next-of-kin on the death certificate or to the mother on the birth certificate. The questionnaires elicit additional information about the decedent or the mother and child, and in this way the survey provides a rich supplement to the information on the basic vital record. The sample data are weighted to provide unbiased estimates for the universe of records from which the sample was drawn.

The 1988 National Maternal and Infant Health Survey (NMIHS) is the most recent follow-back survey conducted by NCHS focusing on reproductive health; it included a nationally representative sample of 9,953 live births, 5,332 infant deaths, and 3,309 late fetal deaths. This survey obtained information on socioeconomic and demographic characteristics of mothers, prenatal care, pregnancy history, working history, health status of mother and infant, types and sources of medical care received, and lifestyle characteristics, including maternal smoking, drinking, and drug use.

In 1991 NCHS conducted a Longitudinal Followup (LF) to the 1988 NMIHS by recontacting the mothers in the earlier survey to get information on their children's health during the 2 or 3 years after birth. The LF requested data from the child's medical providers as well. The LF provided information on infant feeding practices, child care, parental employment, and a wide range of information on early childhood health and development. A subsample of NMIHS women who had infant or fetal deaths (1,000 of each) was included in the LF survey to obtain data on subsequent reproductive behavior following infant or fetal loss.

Since 1999, NCHS has been a collaborating agency with the National Center for Education Statistics on the Early Childhood Longitudinal Survey—Birth Cohort (ECLS-B). The ECLS-B examines children's health, development, care, and education during the formative years from birth through kindergarten. Nearly 11,000 children, sampled from birth certificates from across the United States, were included in the ECLS-B study.

The latest National Mortality Followback Survey (NMFS) was conducted using death records for decedents who died in 1993. This survey included special samples of deaths from homicide, suicide, motor vehicle accidents, other accidents, HIV, and certain natural causes. In addition to information on the use of health services in the last year of life, the 1993 NMFS was unique because it included information obtained from the records of medical examiners and coroners. The NMFS also included information on socioeconomic and demographic characteristics of deceased persons, use of and payment for hospitals and institutional care during the last year of life, and various aspects of life style and other factors related to health status. In 2004, the Last Acts Partnership, a Washington, DC, advocacy organization for quality end-of-life care, recommended an ongoing NMFS (Last Acts Partnership, 2004), but to date no financial support has been forthcoming.

## The National Death Index

The National Death Index (NDI) is a central computerized index of identifying death-record information (beginning with 1979 deaths) at NCHS compiled from files submitted by state vital statistics offices. Working with state offices, NCHS established the NDI as a resource to aid epidemiologists and other health and medical investigators with their mortality ascertainment activities. Death records are added to the NDI file annually, approximately 12 months after the end of a particular calendar year. The NDI is available to investigators solely for statistical purposes in medical and health research; it is not accessible to organizations or the general public for legal, administrative, or genealogy purposes.

The NDI file contains a standard set of identifying information on each death to be used in searches of the file to identify and locate death records in the state offices. NDI users are encouraged to submit as many of the following data items as possible for each study subject: first and last name; middle initial; father's surname; Social Security number; month, day, and year of birth; race; sex; marital status; state of residence; and state of birth. Results of NDI searches assist investigators in determining whether persons in their studies have died and, if so, provide the names of the states in which those deaths occurred, the dates of death, and the corresponding death certificate numbers. Investigators can then either make arrangements with the appropriate state offices to obtain copies of death certificates or obtain cause-of-death codes using the NDI Plus service.

### Vital Statistics Training Program

The NCHS vital statistics training program is another important activity involving interaction with state personnel. DVS annually offers 1-week courses in "Vital Statistics: Measurement and Current Analytic Issues," "Vital Statistics: Measurement and Production," and "Vital Statistics Records and Their Administration." In addition, DVS staff has offered several types of courses on the coding and classification of cause-of-death information from death certificates. The participants in these courses are generally employees of state, county, and city registration offices; however, mortality coders from other countries have also been trained in the use of the NCHS Mortality Medical Data System (MMDS). Over the past 20 years, more than 1,000 employees of these offices have taken one or more of these courses.

## NCHS ACTIVITIES IN DATA PRODUCTION AND DISSEMINATION

### Mortality Cause-of-Death Coding

#### Cause-of-Death Classification

Mortality statistics published as part of the NVSS are coded and classified in accordance with World Health Organization (WHO) regulations, which specify that member nations use the current revision of the International Classification of Diseases, ICD-10. ICD-10 not only details disease classification but also provides definitions, tabulation lists, the format of the cause-of-death section of the death certificate, and the rules for selecting the underlying cause of death. It provides the basic guidance used in virtually all countries for cause-of-death classification; the United States began using ICD-10 effective with deaths occurring in 1999.

WHO has provided a mechanism for updating the classification from time to time through the Update and Revision Committee and the Mortality Reference Group. Staff members in DVS are represented on both of these international groups and regularly attend group meetings, as well as the annual meetings of the heads of the WHO Mortality Classification Centers at various locations around the world. As a result of the periodic updates (minor changes are made annually and major changes every 3 years), NCHS publishes updated versions of the ICD-10 tabular list.

## Mortality Medical Data System (MMDS)

Beginning with the implementation of the eighth revision of the ICD in 1968, NCHS developed and employed several interrelated computer systems to automatically select the underlying cause for each death certificate and to produce multiple cause-of-death data. System automation provides the benefit of greater consistency in the application of classification rules while requiring less extensive coder training. Currently, NCHS employs a suite of computer software, known as the MMDS, to code and classify cause-of-death information for most of the death records registered in the United States. The MMDS software is used by most states and many international partners to standardize the coding and classification of death records.

There are three main software applications that comprise the MMDS. SuperMICAR accepts all literal entries of the certifying physician and automatically converts the reported medical conditions into special numeric entity reference numbers (ERNs). The ERN output from SuperMICAR then becomes input to MICAR200 (Medical Indexing, Classification, and Retrieval), which assigns ICD codes to the ERNs for input to the third application, ACME (Automated Classification of Medical Entities), which in turn selects the underlying cause of death according to the rules of the ICD. (Another program called TRANSAX uses the output from MICAR200 to produce multiple cause-of-death data.) Also, SuperMICAR provides a method to retain literal entries in electronic form for quality control and analysis purposes of rare events subsumed under broad ICD codes. At present, the MMDS handles at least 85 percent of all death certificates; professionally trained classifiers code rejects from the MMDS manually.

Not only utilized by NCHS and the states, MMDS is used in its totality by the following countries: Canada (English speaking), England and Wales, Scotland, Ireland, South Africa, and Australia. The following countries are completely dependent on ACME for determining underlying death cause: Sweden, France, Canada (French speaking), Hungary, and Brazil and will use other portions of the system in the near future. NCHS provides training and support with installation of the software as well as systems updating. Trinidad and Italy are in the process of implementing all or portions of the system. Keeping these systems updated for the countries reflecting WHO annual changes has been a major support issue.

NCHS also provides cause-of-death coding classes for our international partners. Over the last 5 years NCHS has trained staff from Isle of Mauritius, Kenya, Trinidad, Switzerland, Estonia, Italy, Poland, Spain, Latvia, Hungary, England, Czech Republic, Slovenia, Austria, and Tanzania.

### Matched Multiple Birth File

DVS staff developed the Matched Multiple Birth File (MMBF) to facilitate an analysis of characteristics of sets of births and fetal deaths in multiple deliveries. The MMBF currently includes six combined years (1995–2000) of data of matched sets of twins, triplets, and quadruplets in live births and fetal deaths. Live-birth records are linked to the corresponding infant death records for babies who died. Because of concerns for confidentiality with respect to small numbers for multiple births, some data fields are suppressed; no geographic identifiers are shown in the public-use version of this file.

## REPORTS AND PUBLICATIONS

DVS statisticians and analysts produce a variety of publications and reports. There are standard reports that are produced annually from the natality and mortality data files that contain official statistics on U.S. births and deaths for a particular year. After the close of each calendar year, preliminary birth and death files are produced. These are the basis for the reports "Births: Preliminary Data for Year XXXX" and "Deaths: Preliminary Data for Year XXXX." Several months later, the files for the data year are closed and finalized. At that point, the reports "Births: Final Data for Year XXXX" and "Deaths: Final Data for Year XXXX" are produced. DVS analysts also produce annual reports on the linked birth/infant-death data set and the fetal and perinatal mortality data. The release of these "final" reports coincides with the release of the public-use data files for these years. Annual reports are also produced on leading causes of death and life expectancy. Recently, staff members have introduced the public to the expanded health data from the 2003 revised birth certificate through an annual report, "Expanded Health Data from the New Birth Certificate." DVS staff members respond to unexpected findings from the annual reports to produce special analyses. A recent example was in response to the unexpected increase in the infant mortality rate in 2002; the report was entitled "Explaining the 2001–02 Infant Mortality Increase: Data from the Linked Birth/Infant Death Data Set." DVS also produces a monthly report providing provisional counts of births, deaths, marriages, and divorces. DVS staff members regularly contribute to the CDC journal *MMWR* (*Morbidity and Mortality Weekly Report*).

In addition to these NCHS reports, DVS analysts have published in a variety of peer-reviewed journals. These articles may be coauthored with other NCHS analysts or with other CDC or federal colleagues or academic collaborators. The journals in which DVS staff have published in the last 2 years include *Pediatrics, Paediatric and Perinatal Epidemiology, International Journal of Health Services, Journal of Infectious Diseases, Seminars in Perinatology, Hispanic Journal of Behavioral Sciences, Journal of Marriage and Family, Fertility and Sterility, Birth: Issues in Perinatal Care, American Journal of Public Health, American Journal of Epidemiology, Demography, Maternal and Child Health Journal, Birth Defects Research: Clinical and Molecular Teratology, International Journal of Andrology, Diabetes Care, Diabetologia, CA: A Cancer Journal for Clinicians*, and *Injury Prevention*. DVS staff have also contributed invited chapters in a variety of books.

## DATA DISSEMINATION

### Releasing Microdata and Compressed Vital Statistics Files

On November 20, 2007, DVS/NCHS released a new policy on the release of and access to vital statistics microdata for births, deaths, fetal deaths, linked birth/infant death, and matched multiple births. Effective with the 2005 data year, NCHS revised its microdata release and access policy to comply with state requirements, laws, and policies. This DVS revised policy reflects the dual goals to make data available as widely as possible while being responsive to concerns about confidentiality.

Researchers can download public-use microdata files for births and linked birth/infant death data sets directly from the NCHS website, or they can be provided on CD-ROM or DVD.[5] Birth, death, fetal death, and linked birth/infant-death public-use microdata files beginning with the 2005 data year contain individual-level vital-event data at the national level only, that is, with no geographic identifiers (no state, county, or city identifiers).[6]

---

[5] For a complete statement of the DVS/NCHS microdata release policy, see http://www.cdc.gov/nchs/about/major/dvs/NCHS_DataRelease.htm. Microdata refers to records for individual cases. Files released by DVS may include a single record for each birth or death, or the file may be "compressed," replacing identical records with a single record and the number of times that record occurs in the file. A compressed file reduces the number of records in the file. Compressed mortality files produced by NCHS list year and county of death, race (white, black, other), cause of death, and sex, use broad age groupings, and therefore do not contain as much detail as single-record microdata files.

[6] Over the years, confidentiality standards have changed for the public release of geographic and date details on vital statistics microdata files. These changes are reflected in the data available in successive time periods, as follows:

- Birth, death, and fetal death public-use microdata files prior to 1989 contain all counties and exact dates (year, month, and day) of birth and death.
- Birth, death and fetal death public-use microdata files for data years 1989–2004 con-

Researchers may request customized microdata files (birth, death, fetal death, and linked birth/infant death) and compressed files (death only) containing geographic detail for all states and counties for those data years with limited (1989–2004) or no (2005 forward) geographic detail in the public-use files. Data for approved projects are provided at no cost.

### Internet-Based Tabulation Query Systems

Data users may also access data using Internet programs to construct their own tabulations of births and deaths with geographic detail subject to population or cell size limitations. Some of these interactive systems allow users to build tables based on microdata; however, only tabulated data are presented to the user. DVS has constructed an interactive tabulation system called VitalStats on the CDC/NCHS website. VitalStats is based on the Beyond 20-20 software package. Users can tabulate, chart, and map natality, fetal death, and linked birth/infant-death data using prebuilt tables. They can also build their own tables based on natality and fetal-death data files. Trend tabulations of natality, fetal mortality, and linked birth/infant-death data by geographic detail at the county level are currently available.[7]

## LOOKING AHEAD: VITAL STATISTICS FOR THE 21ST CENTURY

### Building on the Present

### Automation of Vital Statistics at the Source, State, and National Levels

Despite the importance of the nation's vital statistics system, in many states it remains based on outmoded vital registration practices and systems,

---

tain only geographic identifiers of counties and cities with a population of 100,000 or greater, and no exact dates. For birth, death, and fetal death files, year, month, and day of week (e.g., Monday) are available.
- Linked birth/infant-death public-use microdata files through 2004 contain geographic identifiers only for counties and cities with 250,000 or greater population and no exact dates. Year, month, and day of week (e.g., Monday) of birth/death are available. Beginning 2005 no geographic identifiers will be included on the public-use linked file.
- The Matched Multiple Births File combines data from the 6 years 1995–2000 but excludes all geographic identifiers and exact dates of births and deaths. The file also excludes year, month, and day of week (e.g., Monday). For a description of the file, see http://www.cdc.gov/nchs/r&d/rdc_twin.htm, and for download, ftp://ftp.cdc.gov/pub/Health_Statistics/NCHS/datasets/mmb2/. An earlier version of this file, also available, combines data for 1995–1998.

[7]Interactive systems currently available are VitalStats at http://www.cdc.gov/nchs/VitalStats.htm; WONDER (Wide-ranging ON-line Data for Epidemiological Research) at http://wonder.cdc.gov; WISQARS (Web-based Injury Statistics Query and Reporting System) at http://www.cdc.gov/ncipc/wisqars/; and IRHA (Interactive Reproductive Health Atlas) at http://www.cdc.gov/reproductivehealth/GISAtlas.

a fact that raises concerns about data quality, timeliness, and the lack of real-time linkage capabilities for the more than 6 million annual vital events. To resolve these issues, vital registration requires more complete automation at the level of primary data collection and changes in the relationships among the providers of source records, the state registration offices, and NCHS. Collection of the cause-of-death information continues to be primarily a paper-based process, unchanged at the local and state levels for the last half century, and the reporting of fetal deaths is no better, if not worse. Although the states have been successful working with the funeral directors in automating the collection of demographic information on the decedent, there have been few successes in automation of the information collected from attending physicians, medical examiners, and coroners who provide and certify medical information on cause of death. The complete death-record data do not become computerized until reaching the state vital registration office, sometimes after considerable delay. The lack of automation at the source precludes timely follow-back to improve data quality and does not take advantage of the expanded use of electronic medical records. Even the electronic sharing of information between states and with NCHS is rudimentary.

To address these problems, NAPHSIS, NCHS, and SSA developed a partnership to improve the responsiveness of state vital registration and statistics systems. Their objective was to improve the timeliness, quality, and sustainability of these systems by adopting national, consensus-based standards and guidelines. Although these guidelines have been developed and have been used by some states to reengineer their birth registration systems and the demographic portion of their mortality systems, much remains to be done. Reengineered systems must include efficient methods for capturing data through standardized data collection instruments, coding specifications, query guidelines, and definitions and transmit these data using HL-7 standardized messaging. As the Nationwide Health Information Network (NHIN) is knitted together, these reengineered vital statistics systems will need to be integrated with other electronic public health systems collecting information on immunizations, newborn screening, and hearing screening, and with electronic health records used by data providers, including hospitals and physicians.

Many questions are yet to be answered. What is the most effective way to retrieve quality medical information from the attending physician, coroner, or medical examiner? How can funeral directors and physicians be connected electronically and share with the state confidential information about the decedent in a secure environment? At what level of specificity do prompts and data edits for the medical information obtained from the physician become counterproductive? Efforts are currently under way to address some of these issues. SSA has been able to provide funding to some states to automate the reporting of the fact of death, and NCHS is working to develop vital statistics data transmission standards. NAPHSIS is developing a data transfer system, yet the most daunting challenge remains the funding of the implementation of these new systems by the states. NCHS has had problems with funding the basic VSCP program and has been of little help in supporting states in their automation needs. As with the states, NCHS's internal systems also need to be completely reengineered to be able to interact with state systems on a real-time basis in order to follow back immediately to improve data quality as well as to publish and provide national vital statistics quickly on a year-to-date basis (Rothwell, 2004). Work is now under way to reengineer the internal systems within DVS to improve data quality and timeliness.

### Follow-back Surveys for Vital Statistics

An Institute of Medicine and National Research Council (2003) report, *Describing Death in America: What We Need to Know*, "highlighted how little we know about 'the quality, appropriateness or costs of care dying individuals receive, or the burden on caregivers and survivors." A reinstitution of the U.S. National Mortality Follow-back Survey (NMFS) could provide the information we must have if we are to improve care and plan intelligently for the future health care needs of our aging population. Such an ongoing follow-back survey, taking advantage of new electronic health records and improved and linked vital statistics systems, could also on a regular basis target causes of death of public health interest and more fully illustrate demographic disparities in mortality.

### Possibilities and Challenges

The automation of vital statistics data collection at the source and its linkage to electronic health records and the building of responsive electronic transmission and linkage systems that will be required by the IRTPA legislation can provide significant new possibilities for how vital statistics are collected and utilized. For example, the 2003 revision of the birth record could be considered a rather extensive perinatal record. In the new environment of automated and linked vital records and electronic health records, should all this information be collected for every record or should only a core of information be collected for each event augmented by a continuous follow-back survey or surveillance capability to sample electronic health records for information needed to elucidate certain health outcomes and health care practices? However, if birth certificates do not require this information, will medical records contain the data in a standardized fashion, useful for sampling? With real-time access to mortality data and linkage to electronic health records, is quicker annual reporting and quicker provision of annual mortality data files the primary outcome, or should this expanded capability be used to once again turn vital statistics into a dynamic public health surveillance system? To help with surveillance, could there be a provisional record collected followed by a more robust reporting of the causes of death for all or selected records? With the future dependence on electronic health records and the growing need to depict chronic conditions, the use of the concept of the underlying cause of death may need to be revisited along with the automated medical coding systems, which turn literal medical nomenclature into an ICD code(s).

The determination of appropriate contributions to a system that supports a variety of government agencies is always difficult, and certainly this has been the case for the support of this nation's vital statistics system. Complicating this situation is that although NCHS is authorized to collect vital statistics, states are not required to provide this information. If vital events are indeed vital should they not be required to be reported? Should a core data set be defined and required for each vital event and be made a reportable event for states to provide to NCHS and then NCHS and its federal partners pay for follow-back on specific records of interest as well as training and maintenance of systems that support data sharing?

Efforts to rejuvenate the nation's vital statistics system are encouraging, but they will need to expand dramatically to provide a timely, high-quality, and flexible system to monitor vital health outcomes at the local, state, and national levels.

# – C –

# Workshop Agenda and Participant List

### AGENDA

Committee on National Statistics
Workshop on Vital Data for National Needs

April 30, 2008
Lecture Room, National Academy of Sciences Building
Washington, DC

**Welcome and Introductory Comments**

8:00 am    Welcome and Introductions
              Louise Ryan, *Planning Committee Chair*

8:30         Background, Workshop Goals, and Agenda
              Constance Citro, *Director, Committee on National Statistics*
              Edward Sondik, *Director, National Center for Health Statistics*
              Louise Ryan, *Planning Committee Chair*

## Uses of Vital Statistics Data and Increasing Demands on the System

8:45      Health Policy and Health Research Uses of Vital Statistics: Data Driven Policy; Health Risk Assessments; Health Surveillance; Health Disparities
Moderator: Louise Ryan, *Department of Biostatistics, Harvard University*
Presenters:
- Nancy Krieger, *Department of Society, Human Development, and Health, Harvard School of Public Health*
- Richard Rogers, *Department of Sociology and Program on Population Processes, University of Colorado, Boulder*
- Peter van Dyck, *Associate Administrator, Maternal and Child Health, Health Resources and Services Administration, U.S. Department of Health and Human Services*

10:00      Break

10:15      Understanding the Future: Life Expectancy, Population, and Fiscal Projections
Moderator: Samuel Preston, *Population Studies Center, University of Pennsylvania*
Presenters:
- Stephen Goss, *Chief Actuary, U.S. Social Security Administration*
- Fred Hollmann, *Population Projections, U.S. Census Bureau*

11:00      Growing and Emerging Uses: National Security, Infectious Disease Surveillance, Small-Area Estimates for Local Planning
Moderator: Kenneth Prewitt, *School of International and Public Affairs, Columbia University*
Presenters:
- Michael A. Stoto, *School of Nursing and Health Studies, Georgetown University*
- Ed Hunter, *Deputy Director, Washington Office, Centers for Disease Control and Prevention*
- Victoria Velkoff, *Assistant Division Chief, Population Estimates and Projections, U.S. Census Bureau*

12:00 pm      Working Lunch and Luncheon Speaker: Reflections on Current Uses and Future Needs for Vital Statistics
Steven Murdock, *Director, U.S. Census Bureau*

## Where Are We Now: Challenges and Needs

1:00      Role of the States and Recent Innovations
Moderator: Edward Perrin, School of Public Health and Community Medicine, University of Washington; *Dr. Perrin was unable to attend and Constance Citro, Committee on National Statistics, substituted as moderator*
Presenters:
- Steven Schwartz, *Registrar and Assistant Commissioner, Office of Vital Statistics, New York City Department of Health and Mental Hygiene*
- Garland Land, *Executive Director, National Association for Public Health Statistics and Information Systems*

2:00      Methodological Issues: Race, Variations in Certificates, Timeliness, Etc.
Moderator: Sharon Arnold, *Vice President, AcademyHealth*
Presenters:
- Jim Weed, *Deputy Director of the Division of Vital Statistics, National Center for Health Statistics, retired*
- Stephanie Ventura, *Chief, Reproductive Statistics Branch, National Center for Health Statistics*
- Robert Anderson, *Chief, Mortality Statistics Branch, National Center for Health Statistics*

2:45      Break

## Options for Building a Vital Statistics System for the 21st Century

3:00      Federal-State Cooperative Systems: Examples of Successes from Within the Federal Statistical System
Moderator: Janet Norwood, *Consultant*
Presenters:
- John Galvin, *Associate Commissioner for Employment Statistics, Bureau of Labor Statistics*
- Andrew White, *Special Assistant to the Commissioner, National Center for Education Statistics*
- Harry Rosenberg, *National Center for Health Statistics, retired*

4:00      Where Do We Go from Here? Band-Aid to Cadillac Solutions
Moderator: Kenneth Prewitt, *Columbia University*
Presenter: Jennifer Madans, *Associate Director for Science, National Center for Health Statistics*

Discussants:
- Howard Hogan, *Associate Director for Demographic Programs, U.S. Census Bureau*
- Steven Schwartz, *New York City Department of Health and Mental Hygiene*
- Nancy Krieger, *Harvard School of Public Health*

5:15  Concluding Remarks
Charles Rothwell, *Director, Division of Vital Statistics, National Center for Health Statistics*
Louise Ryan, *Planning Committee Chair*

5:30  Adjourn

## PARTICIPANT LIST

Robert Anderson, *National Center for Health Statistics*
Sandra Arévalo, *Northeastern University*
Sharon Arnold, *AcademyHealth*
Jana Asher, *Unaffiliated Scholar*
Delton Atkinson, *National Center for Health Statistics*
Heibatollah Baghi, *George Mason University*
Clifton Bailey, *George Mason University*
Patty Becker, *APB Associates*
Amy Branum, *National Center for Health Statistics*
Genet Burka, *District of Columbia Department of Health*
Jo Amato Burns, *Pension Benefit Guarantee Corporation*
Mark Bye, *Social Security Administration*
Virginia Cain, *National Center for Health Statistics*
Carlos Castillo-Salgado, *Pan American Health Organization*
Constance Citro, *Committee on National Statistics*
Lin (Limin) Clegg, *Department of Veterans Affairs*
Mark Denbaly, *Economic Research Service*
Irma Elo, *University of Pennsylvania*
Suzann Evinger, *Office of Management and Budget*
Ron Fecso, *Government Accountability Office*
Christopher Fulcher, *University of Missouri*
Carolyn Fuqua, *National Opinion Research Council*
John Galvin, *Bureau of Labor Statistics*
Andrea Gerger, *Pan American Health Organization*
Alejandro Giusti, *World Health Organization*
Stephen Goss, *Social Security Administration*

Marjorie Greenberg, *National Center for Health Statistics*
Brady Hamilton, *National Center for Health Statistics*
Elizabeth Hamilton, *National Institute on Aging*
Mary Jo Hoeksema, *Population Association of America*
Howard Hogan, *Census Bureau*
Fred Hollmann, *Census Bureau*
Julia Holmes, *National Center for Health Statistics*
Emily Holubowich, *Coalition for Health Services Research*
Isabelle Horon, *Maryland Department of Health*
Ed Hunter, *Centers for Disease Control and Prevention*
David Johnson, *Census Bureau*
Hormuzd Katki, *National Cancer Institute*
Sharon Kirmeyer, *National Center for Health Statistics*
Nancy Krieger, *Harvard School of Public Health*
Caryn Kuebler, *Government Accountability Office*
Garland Land, *National Association for Public Health Statistics and Information Systems*
Jin Hee Lee, *New York Lawyers for the Public Interest*
Linda Loubert, *Morgan State University*
Marian MacDorman, *National Center for Health Statistics*
Jennifer Madans, *National Center for Health Statistics*
Fatima Marinho, *Pan American Health Organization*
Joyce Martin, *National Center for Health Statistics*
Patricia Martin, *Social Security Administration*
Shelly Martinez, *Office of Management and Budget*
TJ Mathews, *National Center for Health Statistics*
Koren Melfi, *Altarum*
Linda Mellgren, *Department of Health and Human Services*
Fay Menacker, *National Center for Health Statistics*
Pauline Mendola, *National Center for Health Statistics*
Mary Moien, *National Center for Health Statistics*
Michael Molla, *National Center for Health Statistics*
Oscar Mujica, *Pan American Health Organization*
Wolfgang Munar, *Inter-American Development Bank*
Steven Murdock, *Census Bureau*
Janet Norwood, *Independent Consultant*
William O'Hare, *Casey Institute*
Frank Olken, *National Science Foundation*
Sarah Orndorff, *George Washington University*
Jennifer Park, *National Center for Education Statistics*
Edward Perrin, *University of Washington*
Patricia Potrzebowski, *Pennsylvania Department of Health*
Samuel Preston, *University of Pennsylvania*

Kenneth Prewitt, *Columbia University*
Richard Rogers, *University of Colorado*
Cynthia Ronzio, *George Washington University*
Harry Rosenberg, *National Center for Health Statistics (ret.)*
Charles Rothwell, *National Center for Health Statistics*
Kara Ryan, *National Council of La Raza*
Louise Ryan, *Harvard University*
Rama Sastry, *Energy Department*
Margo Schwab, *Office of Management and Budget*
Steven Schwartz, *New York City Department of Health and Mental Hygiene*
Michael Sellner, *Census Bureau*
Jacob Siegel, *Georgetown University*
Howard Silver, *Consortium of Social Science Associations*
Michael Simpson, *Congressional Budget Office*
Michael Siri, *CNSTAT*
Monroe Sirken, *National Center for Health Statistics*
Robyn Sneeringer, *Altarum Institute*
Ed Sondik, *National Center for Health Statistics*
Ed Spar, *Council of Professional Associations on Federal Statistics*
Michael Stoto, *Georgetown University*
Paul Sutton, *National Center for Health Statistics*
Peter van Dyck, *Department of Health and Human Services*
Victoria Velkoff, *Census Bureau*
Stephanie Ventura, *National Center for Health Statistics*
Alice Wade, *Social Security Administration*
Katherine Wallman, *Office of Management and Budget*
James Weed, *National Center for Health Statistics (ret.)*
Rob Weinzimer, *National Center for Health Statistics*
Andrew White, *National Center for Education Statistics*
Al Winters, *Social Security Administration*
Karen Woodrow-Lafield, *Unaffiliated Scholar*

– D –

# 2003 Revisions, Standard Certificates of Death and Live Birth

This appendix reproduces the standard certificates of death and of live birth as of the 2003 round of revisions. Electronic files of the form of the certificates and additional details and instructions are available at http://www.cdc.gov/nchs/vital_certs_rev.htm.

# U.S. STANDARD CERTIFICATE OF DEATH

LOCAL FILE NO. _____  STATE FILE NO. _____

**NAME OF DECEDENT** — *To Be Completed/ Verified By: FUNERAL DIRECTOR:*

1. DECEDENT'S LEGAL NAME (Include AKA's if any) (First, Middle, Last)
2. SEX
3. SOCIAL SECURITY NUMBER

4a. AGE-Last Birthday (Years) | 4b. UNDER 1 YEAR (Months / Days) | 4c. UNDER 1 DAY (Hours / Minutes) | 5. DATE OF BIRTH (Mo/Day/Yr) | 6. BIRTHPLACE (City and State or Foreign Country)

7a. RESIDENCE-STATE | 7b. COUNTY | 7c. CITY OR TOWN

7d. STREET AND NUMBER | 7e. APT. NO. | 7f. ZIP CODE | 7g. INSIDE CITY LIMITS? ☐ Yes ☐ No

8. EVER IN US ARMED FORCES? ☐ Yes ☐ No
9. MARITAL STATUS AT TIME OF DEATH ☐ Married ☐ Married, but separated ☐ Widowed ☐ Divorced ☐ Never Married ☐ Unknown
10. SURVIVING SPOUSE'S NAME (If wife, give name prior to first marriage)

11. FATHER'S NAME (First, Middle, Last)
12. MOTHER'S NAME PRIOR TO FIRST MARRIAGE (First, Middle, Last)

13a. INFORMANT'S NAME | 13b. RELATIONSHIP TO DECEDENT | 13c. MAILING ADDRESS (Street and Number, City, State, Zip Code)

*To Be Completed/ Verified By: FUNERAL DIRECTOR:*

14. PLACE OF DEATH (Check only one: see instructions)

IF DEATH OCCURRED IN A HOSPITAL: ☐ Inpatient ☐ Emergency Room/Outpatient ☐ Dead on Arrival
IF DEATH OCCURRED SOMEWHERE OTHER THAN A HOSPITAL: ☐ Hospice facility ☐ Nursing home/Long term care facility ☐ Decedent's home ☐ Other (Specify)

15. FACILITY NAME (If not institution, give street & number) | 16. CITY OR TOWN , STATE, AND ZIP CODE | 17. COUNTY OF DEATH

18. METHOD OF DISPOSITION: ☐ Burial ☐ Cremation ☐ Donation ☐ Entombment ☐ Removal from State ☐ Other (Specify)
19. PLACE OF DISPOSITION (Name of cemetery, crematory, other place)

20. LOCATION-CITY, TOWN, AND STATE
21. NAME AND COMPLETE ADDRESS OF FUNERAL FACILITY

22. SIGNATURE OF FUNERAL SERVICE LICENSEE OR OTHER AGENT
23. LICENSE NUMBER (Of Licensee)

**ITEMS 24-28 MUST BE COMPLETED BY PERSON WHO PRONOUNCES OR CERTIFIES DEATH**

24. DATE PRONOUNCED DEAD (Mo/Day/Yr)
25. TIME PRONOUNCED DEAD
26. SIGNATURE OF PERSON PRONOUNCING DEATH (Only when applicable)
27. LICENSE NUMBER
28. DATE SIGNED (Mo/Day/Yr)

29. ACTUAL OR PRESUMED DATE OF DEATH (Mo/Day/Yr) (Spell Month)
30. ACTUAL OR PRESUMED TIME OF DEATH
31. WAS MEDICAL EXAMINER OR CORONER CONTACTED? ☐ Yes ☐ No

**CAUSE OF DEATH (See instructions and examples)**

32. PART I. Enter the chain of events--diseases, injuries, or complications--that directly caused the death. DO NOT enter terminal events such as cardiac arrest, respiratory arrest, or ventricular fibrillation without showing the etiology. DO NOT ABBREVIATE. Enter only one cause on a line. Add additional lines if necessary.

Approximate interval: Onset to death

IMMEDIATE CAUSE (Final disease or condition resulting in death) → a. _____
Due to (or as a consequence of):

Sequentially list conditions, if any, leading to the cause listed on line a. Enter the UNDERLYING CAUSE (disease or injury that initiated the events resulting in death) LAST

b. _____
Due to (or as a consequence of):

c. _____
Due to (or as a consequence of):

d. _____

PART II. Enter other significant conditions contributing to death but not resulting in the underlying cause given in PART I

33. WAS AN AUTOPSY PERFORMED? ☐ Yes ☐ No
34. WERE AUTOPSY FINDINGS AVAILABLE TO COMPLETE THE CAUSE OF DEATH? ☐ Yes ☐ No

*To Be Completed By: MEDICAL CERTIFIER*

35. DID TOBACCO USE CONTRIBUTE TO DEATH? ☐ Yes ☐ Probably ☐ No ☐ Unknown
36. IF FEMALE: ☐ Not pregnant within past year ☐ Pregnant at time of death ☐ Not pregnant, but pregnant within 42 days of death ☐ Not pregnant, but pregnant 43 days to 1 year before death ☐ Unknown if pregnant within the past year
37. MANNER OF DEATH ☐ Natural ☐ Homicide ☐ Accident ☐ Pending Investigation ☐ Suicide ☐ Could not be determined

38. DATE OF INJURY (Mo/Day/Yr) (Spell Month)
39. TIME OF INJURY
40. PLACE OF INJURY (e.g., Decedent's home; construction site; restaurant; wooded area)
41. INJURY AT WORK? ☐ Yes ☐ No

42. LOCATION OF INJURY: State: ___ City or Town: ___ Street & Number: ___ Apartment No.: ___ Zip Code: ___

43. DESCRIBE HOW INJURY OCCURRED:

44. IF TRANSPORTATION INJURY, SPECIFY: ☐ Driver/Operator ☐ Passenger ☐ Pedestrian ☐ Other (Specify)

45. CERTIFIER (Check only one):
☐ Certifying physician-To the best of my knowledge, death occurred due to the cause(s) and manner stated.
☐ Pronouncing & Certifying physician-To the best of my knowledge, death occurred at the time, date, and due to the cause(s) and manner stated.
☐ Medical Examiner/Coroner-On the basis of examination, and/or investigation, in my opinion, death occurred at the time, date, and place, and due to the cause(s) and manner stated.

Signature of certifier: _____

46. NAME, ADDRESS, AND ZIP CODE OF PERSON COMPLETING CAUSE OF DEATH (Item 32)

47. TITLE OF CERTIFIER | 48. LICENSE NUMBER | 49. DATE CERTIFIED (Mo/Day/Yr) | 50. **FOR REGISTRAR ONLY**- DATE FILED (Mo/Day/Yr)

*To Be Completed By: FUNERAL DIRECTOR*

51. DECEDENT'S EDUCATION-Check the box that best describes the highest degree or level of school completed at the time of death.
☐ 8th grade or less
☐ 9th - 12th grade; no diploma
☐ High school graduate or GED completed
☐ Some college credit, but no degree
☐ Associate degree (e.g., AA, AS)
☐ Bachelor's degree (e.g., BA, AB, BS)
☐ Master's degree (e.g., MA, MS, MEng, MEd, MSW, MBA)
☐ Doctorate (e.g., PhD, EdD) or Professional degree (e.g., MD, DDS, DVM, LLB, JD)

52. DECEDENT OF HISPANIC ORIGIN? Check the box that best describes whether the decedent is Spanish/Hispanic/Latino. Check the 'No' box if decedent is not Spanish/Hispanic/Latino.
☐ No, not Spanish/Hispanic/Latino
☐ Yes, Mexican, Mexican American, Chicano
☐ Yes, Puerto Rican
☐ Yes, Cuban
☐ Yes, other Spanish/Hispanic/Latino (Specify) _____

53. DECEDENT'S RACE (Check one or more races to indicate what the decedent considered himself or herself to be)
☐ White
☐ Black or African American
☐ American Indian or Alaska Native (Name of the enrolled or principal tribe) _____
☐ Asian Indian
☐ Chinese
☐ Filipino
☐ Japanese
☐ Korean
☐ Vietnamese
☐ Other Asian (Specify) _____
☐ Native Hawaiian
☐ Guamanian or Chamorro
☐ Samoan
☐ Other Pacific Islander (Specify) _____
☐ Other (Specify) _____

54. DECEDENT'S USUAL OCCUPATION (Indicate type of work done during most of working life. DO NOT USE RETIRED).

55. KIND OF BUSINESS/INDUSTRY

REV. 11/2003

# APPENDIX D

## MEDICAL CERTIFIER INSTRUCTIONS for selected items on U.S. Standard Certificate of Death
(See Physicians' Handbook or Medical Examiner/Coroner Handbook on Death Registration for instructions on all items)

### ITEMS ON WHEN DEATH OCCURRED
Items 24-25 and 29-31 should always be completed. If the facility uses a separate pronouncer or other person to indicate that death has taken place with another person more familiar with the case completing the remainder of the medical portion of the death certificate, the pronouncer completes Items 24-28. If a certifier completes Items 24-25 as well as Items 29-49, Items 26-28 may be left blank.

### ITEMS 24-25, 29-30 – DATE AND TIME OF DEATH
Spell out the name of the month. If the exact date of death is unknown, enter the **approximate** date. If the date cannot be approximated, enter the date the body is found and identify as **date found**. Date pronounced and actual date may be the same. Enter the exact hour and minutes according to a 24-hour clock; estimates may be provided with "Approx." placed before the time.

### ITEM 32 – CAUSE OF DEATH (See attached examples)
Take care to make the entry legible. Use a computer printer with high resolution, typewriter with good black ribbon and clean keys, or print legibly using permanent **black** ink in completing the CAUSE OF DEATH Section. **Do not abbreviate** conditions entered in section.

**Part I** (Chain of events leading directly to death)
- Only **one** cause should be entered on each line. Line (a) **MUST ALWAYS** have an entry. **DO NOT** leave blank. Additional lines may be added if necessary.
- If the condition on Line (a) resulted from an underlying condition, put the underlying condition on Line (b), and so on, until the full sequence is reported. **ALWAYS** enter the **underlying cause of death** on the lowest used line in Part I.
- For each cause indicate the best estimate of the interval between the presumed onset and the date of death. The terms "unknown" or "approximately" may be used. General terms, such as minutes, hours, or days, are acceptable, if necessary. **DO NOT** leave blank.
- The terminal event (for example, cardiac arrest or respiratory arrest) should not be used. If a mechanism of death seems most appropriate to you for line (a), then you must always list its cause(s) on the line(s) below it (for example, cardiac arrest **due to** coronary artery atherosclerosis or cardiac arrest **due to** blunt impact to chest).
- If an organ system failure such as congestive heart failure, hepatic failure, renal failure, or respiratory failure is listed as a cause of death, always report its etiology on the line(s) beneath it (for example, renal failure **due to** Type I diabetes mellitus).
- When indicating neoplasms as a cause of death, include the following: 1) primary site or that the primary site is unknown, 2) benign or malignant, 3) cell type *or* that the cell type is unknown, 4) grade of neoplasm, and 5) part or lobe of organ affected. (For example, a primary well-differentiated squamous cell carcinoma, lung, left upper lobe.)
- Always report the fatal injury (for example, stab wound of chest), the trauma (for example, transection of subclavian vein), and impairment of function (for example, air embolism).

**PART II** (Other significant conditions)
- Enter all diseases or conditions contributing to death that were not reported in the chain of events in Part I and that did not result in the **underlying cause of death**. See attached examples.
- If two or more possible sequences resulted in death, or if two conditions seem to have added together, report in Part I the one that, in your opinion, most directly caused death. Report in Part II the other conditions or diseases.

### CHANGES TO CAUSE OF DEATH
Should additional medical information or autopsy findings become available that would change the cause of death originally reported, the original death certificate should be amended by the certifying physician by **immediately** reporting the revised cause of death to the State Vital Records Office.

### ITEMS 33-34 - AUTOPSY
- 33 - Enter "Yes" if either a partial or full autopsy was performed. Otherwise enter "No."
- 34 - Enter "Yes" if autopsy findings were available to complete the cause of death; otherwise enter "No". Leave item blank if no autopsy was performed.

### ITEM 35 - DID TOBACCO USE CONTRIBUTE TO DEATH?
Check "yes" if, in your opinion, the use of tobacco contributed to death. Tobacco use may contribute to deaths due to a wide variety of diseases; for example, tobacco use contributes to many deaths due to emphysema or lung cancer and some heart disease and cancers of the head and neck. Check "no" if, in your clinical judgment, tobacco use did not contribute to this particular death.

### ITEM 36 - IF FEMALE, WAS DECEDENT PREGNANT AT TIME OF DEATH OR WITHIN PAST YEAR?
*This information is important in determining pregnancy-related mortality.*

### ITEM 37 - MANNER OF DEATH
- Always check Manner of Death, which is important: 1) in determining accurate causes of death; 2) in processing insurance claims; and 3) in statistical studies of injuries and death.
- Indicate "Pending investigation" if the manner of death cannot be determined whether due to an accident, suicide, or homicide within the statutory time limit for filing the death certificate. This should be changed later to one of the other terms.
- Indicate "Could not be Determined" **ONLY** when it is impossible to determine the manner of death.

### ITEMS 38-44 - ACCIDENT OR INJURY – to be filled out in all cases of deaths due to injury or poisoning.
- 38 - Enter the exact month, day, and year of injury. Spell out the name of the month. **DO NOT** use a number for the month. (Remember, the date of injury may differ from the date of death.) Estimates may be provided with "Approx." placed before the date.
- 39 - Enter the exact hour and minutes of injury or use your best estimate. Use a 24-hour clock.
- 40 - Enter the general place (such as restaurant, vacant lot, or home) where the injury occurred. **DO NOT** enter firm or organization names. (For example, enter "factory", **not** "Standard Manufacturing, Inc.".)
- 41 - Complete if anything other than natural disease is mentioned in Part I or Part II of the medical certification, including homicides, suicides, and accidents. This includes all motor vehicle deaths. The item **must** be completed for decedents ages 14 years or over and may be completed for those less than 14 years of age if warranted. Enter "Yes" if the injury occurred at work. Otherwise enter "No". An injury may occur at work regardless of whether the injury occurred in the course of the decedent's "usual" occupation. Examples of injury at work and injury not at work follow:

**Injury at work**
Injury while working or in vocational training on job premises
Injury while on break or at lunch or in parking lot on job premises
Injury while working for pay or compensation, including at home
Injury while working as a volunteer law enforcement official etc.
Injury while traveling on business, including to/from business contacts

**Injury not at work**
Injury while engaged in personal recreational activity on job premises
Injury while a visitor (not on official work business) to job premises
Homemaker working at homemaking activities
Student in school
Working for self for no profit (mowing yard, repairing own roof, hobby)
Commuting to or from work

- 42 - Enter the complete address where the injury occurred including zip code.
- 43 - Enter a brief but specific and clear description of how the injury occurred. Explain the circumstances or cause of the injury. Specify **type of gun** or **type of vehicle** (e.g., car, bulldozer, train, etc.) when relevant to circumstances. Indicate if more than one vehicle involved; specify type of vehicle decedent was in.
- 44 - Specify role of decedent (e.g. driver, passenger). Driver/operator and passenger should be designated for modes other than motor vehicles such as bicycles. Other applies to watercraft, aircraft, animal, or people attached to outside of vehicles (e.g. surfers).

**Rationale:** Motor vehicle accidents are a major cause of unintentional deaths; details will help determine effectiveness of current safety features and laws.

**REFERENCES**
For more information on how to complete the medical certification section of the death certificate, refer to tutorial at http://www.TheNAME.org and resources including instructions and handbooks available by request from NCHS, Room 7318, 3311 Toledo Road, Hyattsville, Maryland 20782-2003 or at www.cdc.gov/nchs/about/major/dvs/handbk.htm

REV. 11/2003

## Cause-of-death – Background, Examples, and Common Problems

Accurate cause of death information is important
• to the public health community in evaluating and improving the health of all citizens, and
• often to the family, now and in the future, and to the person settling the decedent's estate.

The cause-of-death section consists of two parts. **Part I** is for reporting a chain of events leading directly to death, with the **immediate cause** of death (the final disease, injury, or complication directly causing death) on line a and the **underlying cause** of death (the disease or injury that initiated the chain of events that led directly and inevitably to death) on the lowest used line. **Part II** is for reporting all other significant diseases, conditions, or injuries that contributed to death but which did not result in the underlying cause of death given in Part I. The cause-of-death information should be YOUR best medical OPINION. A condition can be listed as "probable" even if it has not been definitively diagnosed.

### Examples of properly completed medical certifications

**CAUSE OF DEATH (See instructions and examples)**

32. PART I. Enter the chain of events–diseases, injuries, or complications–that directly caused the death. DO NOT enter terminal events such as cardiac arrest, respiratory arrest, or ventricular fibrillation without showing the etiology. DO NOT ABBREVIATE. Enter only one cause on a line. Add additional lines if necessary.

Approximate interval: Onset to death

IMMEDIATE CAUSE (Final disease or condition resulting in death) → a. **Rupture of myocardium** — Minutes
Due to (or as a consequence of):

Sequentially list conditions, if any, leading to the cause listed on line a. Enter the UNDERLYING CAUSE (disease or injury that initiated the events resulting in death) LAST

b. **Acute myocardial infarction** — 6 days
Due to (or as a consequence of):

c. **Coronary artery thrombosis** — 5 years
Due to (or as a consequence of):

d. **Atherosclerotic coronary artery disease** — 7 years

PART II. Enter other significant conditions contributing to death but not resulting in the underlying cause given in PART I
**Diabetes, Chronic obstructive pulmonary disease, smoking**

33. WAS AN AUTOPSY PERFORMED? ■ Yes □ No
34. WERE AUTOPSY FINDINGS AVAILABLE TO COMPLETE THE CAUSE OF DEATH? ■ Yes □ No

35. DID TOBACCO USE CONTRIBUTE TO DEATH?
■ Yes □ Probably
□ No □ Unknown

36. IF FEMALE:
□ Not pregnant within past year
□ Pregnant at time of death
□ Not pregnant, but pregnant within 42 days of death
□ Not pregnant, but pregnant 43 days to 1 year before death
□ Unknown if pregnant within the past year

37. MANNER OF DEATH
■ Natural □ Homicide
□ Accident □ Pending Investigation
□ Suicide □ Could not be determined

---

**CAUSE OF DEATH (See instructions and examples)**

32. PART I. Enter the chain of events–diseases, injuries, or complications–that directly caused the death. DO NOT enter terminal events such as cardiac arrest, respiratory arrest, or ventricular fibrillation without showing the etiology. DO NOT ABBREVIATE. Enter only one cause on a line. Add additional lines if necessary.

Approximate interval: Onset to death

IMMEDIATE CAUSE (Final disease or condition resulting in death) → a. **Aspiration pneumonia** — 2 Days
Due to (or as a consequence of):

Sequentially list conditions, if any, leading to the cause listed on line a. Enter the UNDERLYING CAUSE (disease or injury that initiated the events resulting in death) LAST

b. **Complications of coma** — 7 weeks
Due to (or as a consequence of):

c. **Blunt force injuries** — 7 weeks
Due to (or as a consequence of):

d. **Motor vehicle accident** — 7 weeks

PART II. Enter other significant conditions contributing to death but not resulting in the underlying cause given in PART I

33. WAS AN AUTOPSY PERFORMED? ■ Yes □ No
34. WERE AUTOPSY FINDINGS AVAILABLE TO COMPLETE THE CAUSE OF DEATH? ■ Yes □ No

35. DID TOBACCO USE CONTRIBUTE TO DEATH?
□ Yes □ Probably
■ No □ Unknown

36. IF FEMALE:
□ Not pregnant within past year
□ Pregnant at time of death
□ Not pregnant, but pregnant within 42 days of death
□ Not pregnant, but pregnant 43 days to 1 year before death
□ Unknown if pregnant within the past year

37. MANNER OF DEATH
□ Natural □ Homicide
■ Accident □ Pending Investigation
□ Suicide □ Could not be determined

38. DATE OF INJURY (Mo/Day/Yr) (Spell Month): August 15, 2003
39. TIME OF INJURY: Approx. 2320
40. PLACE OF INJURY (e.g., Decedent's home; construction site; restaurant; wooded area): road side near state highway
41. INJURY AT WORK? □ Yes ■ No

42. LOCATION OF INJURY: State: Missouri   City or Town: near Alexandria
Street & Number: mile marker 17 on state route 46a   Apartment No.:   Zip Code:

43. DESCRIBE HOW INJURY OCCURRED:
**Decedent driver of van, ran off road into tree**

44. IF TRANSPORTATION INJURY, SPECIFY:
■ Driver/Operator
□ Passenger
□ Pedestrian
□ Other (Specify)

### Common problems in death certification

The **elderly decedent** should have a clear and distinct etiological sequence for cause of death, if possible. Terms such as senescence, infirmity, old age, and advanced age have little value for public health or medical research. Age is recorded elsewhere on the certificate. When a number of conditions resulted in death, the physician should choose the single sequence that, in his or her opinion, best describes the process leading to death, and place any other pertinent conditions in Part II. If after careful consideration the physician cannot determine a sequence that ends in death, then the medical examiner or coroner should be consulted about conducting an investigation or providing assistance in completing the cause of death.

The **infant decedent** should have a clear and distinct etiological sequence for cause of death, if possible. "Prematurity" should not be entered without explaining the etiology of prematurity. Maternal conditions may have initiated or affected the sequence that resulted in infant death, and such maternal causes should be reported in addition to the infant causes on the infant's death certificate (e.g., Hyaline membrane disease due to prematurity, 28 weeks due to placental abruption due to blunt trauma to mother's abdomen).

When **SIDS** is suspected, a complete investigation should be conducted, typically by a medical examiner or coroner. If the infant is under 1 year of age, no cause of death is determined after scene investigation, clinical history is reviewed, and a complete autopsy is performed, then the death can be reported as Sudden Infant Death Syndrome.

When processes such as the following are reported, additional information about the etiology should be reported:

| | | | | |
|---|---|---|---|---|
| Abscess | Carcinomatosis | Disseminated intra vascular coagulopathy | Hyponatremia | Pulmonary arrest |
| Abdominal hemorrhage | Cardiac arrest | Dysrhythmia | Hypotension | Pulmonary edema |
| Adhesions | Cardiac dysrhythmia | End-stage liver disease | Immunosuppression | Pulmonary embolism |
| Adult respiratory distress syndrome | Cardiomyopathy | End-stage renal disease | Increased intra cranial pressure | Pulmonary insufficiency |
| Acute myocardial infarction | Cardiopulmonary arrest | Epidural hematoma | Intra cranial hemorrhage | Renal failure |
| Altered mental status | Cellulitis | Exsanguination | Malnutrition | Respiratory arrest |
| Anemia | Cerebral edema | Failure to thrive | Metabolic encephalopathy | Seizures |
| Anoxia | Cerebrovascular accident | Fracture | Multi-organ failure | Sepsis |
| Anoxic encephalopathy | Cerebellar tonsillar herniation | Gangrene | Multi-system organ failure | Septic shock |
| Arrhythmia | Chronic bedridden state | Gastrointestinal hemorrhage | Myocardial infarction | Shock |
| Ascites | Cirrhosis | Heart failure | Necrotizing soft-tissue infection | Starvation |
| Aspiration | Coagulopathy | Hemothorax | Old age | Subdural hematoma |
| Atrial fibrillation | Compression fracture | Hepatic failure | Open (or closed) head injury | Subarachnoid hemorrhage |
| Bacteremia | Congestive heart failure | Hepatitis | Paralysis | Sudden death |
| Bedridden | Convulsions | Hepatorenal syndrome | Pancytopenia | Thrombocytopenia |
| Biliary obstruction | Decubiti | Hyperglycemia | Perforated gallbladder | Uncal herniation |
| Bowel obstruction | Dehydration | Hyperkalemia | Peritonitis | Urinary tract infection |
| Brain injury | Dementia (when not otherwise specified) | Hypovolemic shock | Pleural effusions | Ventricular fibrillation |
| Brain stem herniation | Diarrhea | | Pneumonia | Ventricular tachycardia |
| Carcinogenesis | | | | Volume depletion |

If the certifier is unable to determine the etiology of a process such as those shown above, the process must be qualified as being of an unknown, undetermined, probable, presumed, or unspecified etiology so it is clear that a distinct etiology was not inadvertently or carelessly omitted.

The following conditions and types of death might seem to be specific or natural but when the medical history is examined further may be found to be complications of an injury or poisoning (possibly occurring long ago). Such cases should be reported to the medical examiner/coroner.

| | | | |
|---|---|---|---|
| Asphyxia | Epidural hematoma | Hip fracture | Pulmonary emboli | Subdural hematoma |
| Bolus | Exsanguination | Hyperthermia | Seizure disorder | Surgery |
| Choking | Fall | Hypothermia | Sepsis | Thermal burns/chemical burns |
| Drug or alcohol overdose/drug or alcohol abuse | Fracture | Open reduction of fracture | Subarachnoid hemorrhage | |

REV. 11/2003

# FUNERAL DIRECTOR INSTRUCTIONS for selected items on U.S.

**Standard Certificate of Death** (For additional information concerning all items on certificate see Funeral Directors' Handbook on Death Registration)

### ITEM 1. DECEDENT'S LEGAL NAME
Include any other names used by decedent, if substantially different from the legal name, after the abbreviation AKA (also known as) e.g. Samuel Langhorne Clemens AKA Mark Twain, **but not** Jonathon Doe AKA John Doe

### ITEM 5. DATE OF BIRTH
Enter the full name of the month (January, February, March etc.) Do not use a number or abbreviation to designate the month.

### ITEM 7A-G. RESIDENCE OF DECEDENT (information divided into seven categories)
Residence of decedent is the place where the decedent actually resided. The place of residence is not necessarily the same as "home state" or "legal residence". Never enter a temporary residence such as one used during a visit, business trip, or vacation. Place of residence during a tour of military duty or during attendance at college is considered permanent and should be entered as the place of residence. If the decedent had been living in a facility where an individual usually resides for a long period of time, such as a group home, mental institution, nursing home, penitentiary, or hospital for the chronically ill, report the location of that facility in item 7. If the decedent was an infant who never resided at home, the place of residence is that of the parent(s) or legal guardian. **Never** use an acute care hospital's location as the place of residence for any infant. If Canadian residence, please specify Province instead of State.

### ITEM 10. SURVIVING SPOUSE'S NAME
If the decedent was married at the time of death, enter the full name of the surviving spouse. If the surviving spouse is the wife, enter her name prior to first marriage. This item is used in establishing proper insurance settlements and other survivor benefits.

### ITEM 12. MOTHER'S NAME PRIOR TO FIRST MARRIAGE
Enter the name used prior to first marriage, commonly known as the maiden name. This name is useful because it remains constant throughout life.

### ITEM 14. PLACE OF DEATH
The place where death is pronounced should be considered the place where death occurred. If the place of death is unknown but the body is found in your State, the certificate of death should be completed and filed in accordance with the laws of your State. Enter the place where the body is found as the place of death.

### ITEM 51. DECEDENT'S EDUCATION *(Check appropriate box on death certificate)*
Check the box that corresponds to the highest level of education that the decedent completed. **Information in this section will not appear on the certified copy of the death certificate.** This information is used to study the relationship between mortality and education (which roughly corresponds with socioeconomic status). This information is valuable in medical studies of causes of death and in programs to prevent illness and death.

### ITEM 52. WAS DECEDENT OF HISPANIC ORIGIN? *(Check "No" or appropriate "Yes" box)*
Check "No" or check the "Yes" box that best corresponds with the decedent's ethnic Spanish identity as given by the informant. Note that "Hispanic" is not a race and item 53 must also be completed. Do not leave this item blank. With respect to this item, "Hispanic" refers to people whose origins are from Spain, Mexico, or the Spanish-speaking Caribbean Islands or countries of Central or South America. Origin includes ancestry, nationality, and lineage. There is no set rule about how many generations are to be taken into account in determining Hispanic origin; it may be based on the country of origin of a parent, grandparent, or some far-removed ancestor. Although the prompts include the major Hispanic groups, other groups may be specified under "other". "Other" may also be used for decedents of multiple Hispanic origin (e.g. Mexican-Puerto Rican). **Information in this section will not appear on the certified copy of the death certificate.** This information is needed to identify health problems in a large minority population in the United States. Identifying health problems will make it possible to target public health resources to this important segment of our population.

### ITEM 53. RACE *(Check appropriate box or boxes on death certificate)*
Enter the race of the decedent as stated by the informant. Hispanic is not a race; information on Hispanic ethnicity is collected separately in item 52. American Indian and Alaska Native refer only to those native to North and South America (including Central America) and does not include Asian Indian. Please specify the name of enrolled or principal tribe (e.g., Navajo, Cheyenne, etc.) for the American Indian or Alaska Native. For Asians check Asian Indian, Chinese, Filipino, Japanese, Korean, Vietnamese, or specify other Asian group; for Pacific Islanders check Guamanian or Chamorro, Samoan, or specify other Pacific Island group. If the decedent was of mixed race, enter each race (e.g., Samoan-Chinese-Filipino or White, American Indian). **Information in this section will not appear on the certified copy of the death certificate. Race is essential for identifying specific mortality patterns and leading causes of death among different racial groups.** It is also used to determine if specific health programs are needed in particular areas and to make population estimates.

### ITEMS 54 AND 55. OCCUPATION AND INDUSTRY
Questions concerning occupation and industry must be completed for all decedents 14 years of age or older. This information is useful in studying deaths related to jobs and in identifying any new risks. For example, the link between lung disease and lung cancer and asbestos exposure in jobs such as shipbuilding or construction was made possible by this sort of information on death certificates. **Information in this section will not appear on the certified copy of the death certificate.**

### ITEM 54. DECEDENT'S USUAL OCCUPATION
Enter the usual occupation of the decedent. This is not necessarily the last occupation of the decedent. Never enter "retired". Give kind of work decedent did during most of his or her working life, such as claim adjuster, farmhand, coal miner, janitor, store manager, college professor, or civil engineer. If the decedent was a homemaker at the time of death but had worked outside the household during his or her working life, enter that occupation. If the decedent was a homemaker during most of his or her working life, and never worked outside the household, enter "homemaker". Enter "student" if the decedent was a student at the time of death and was never regularly employed or employed full time during his or her working life. **Information in this section will not appear on the certified copy of the death certificate.**

### ITEM 55. KIND OF BUSINESS/INDUSTRY
Kind of business to which occupation in item 54 is related, such as insurance, farming, coal mining, hardware store, retail clothing, university, or government. DO NOT enter firm or organization names. If decedent was a homemaker as indicated in item 54, then enter either "own home" or "someone else's home" as appropriate. If decedent was a student as indicated in item 54, then enter type of school, such as high school or college, in item 55. **Information in this section will not appear on the certified copy of the death certificate.**

NOTE: This recommended standard death certificate is the result of an extensive evaluation process. Information on the process and resulting recommendations as well as plans for future activities is available on the Internet at: http://www.cdc.gov/nchs/vital_certs_rev.htm.

REV. 11/2003

## U.S. STANDARD CERTIFICATE OF LIVE BIRTH

**LOCAL FILE NO.** _____  **BIRTH NUMBER:** _____

### CHILD
1. CHILD'S NAME (First, Middle, Last, Suffix)
2. TIME OF BIRTH (24 hr)
3. SEX
4. DATE OF BIRTH (Mo/Day/Yr)
5. FACILITY NAME (If not institution, give street and number)
6. CITY, TOWN, OR LOCATION OF BIRTH
7. COUNTY OF BIRTH

### MOTHER
8a. MOTHER'S CURRENT LEGAL NAME (First, Middle, Last, Suffix)
8b. DATE OF BIRTH (Mo/Day/Yr)
8c. MOTHER'S NAME PRIOR TO FIRST MARRIAGE (First, Middle, Last, Suffix)
8d. BIRTHPLACE (State, Territory, or Foreign Country)
9a. RESIDENCE OF MOTHER-STATE
9b. COUNTY
9c. CITY, TOWN, OR LOCATION
9d. STREET AND NUMBER
9e. APT. NO.
9f. ZIP CODE
9g. INSIDE CITY LIMITS? ☐ Yes ☐ No

### FATHER
10a. FATHER'S CURRENT LEGAL NAME (First, Middle, Last, Suffix)
10b. DATE OF BIRTH (Mo/Day/Yr)
10c. BIRTHPLACE (State, Territory, or Foreign Country)

### CERTIFIER
11. CERTIFIER'S NAME: _____
    TITLE: ☐ MD  ☐ DO  ☐ HOSPITAL ADMIN.  ☐ CNM/CM  ☐ OTHER MIDWIFE
    ☐ OTHER (Specify) _____
12. DATE CERTIFIED __/__/____  MM DD YYYY
13. DATE FILED BY REGISTRAR __/__/____  MM DD YYYY

### INFORMATION FOR ADMINISTRATIVE USE

### MOTHER
14. MOTHER'S MAILING ADDRESS: ☐ Same as residence, or: State: _____ City, Town, or Location:
    Street & Number: _____ Apartment No.: _____ Zip Code: _____
15. MOTHER MARRIED? (At birth, conception, or any time between) ☐ Yes ☐ No
    IF NO, HAS PATERNITY ACKNOWLEDGEMENT BEEN SIGNED IN THE HOSPITAL? ☐ Yes ☐ No
16. SOCIAL SECURITY NUMBER REQUESTED FOR CHILD? ☐ Yes ☐ No
17. FACILITY ID. (NPI)
18. MOTHER'S SOCIAL SECURITY NUMBER:
19. FATHER'S SOCIAL SECURITY NUMBER:

### INFORMATION FOR MEDICAL AND HEALTH PURPOSES ONLY

### MOTHER
20. MOTHER'S EDUCATION (Check the box that best describes the highest degree or level of school completed at the time of delivery)
    ☐ 8th grade or less
    ☐ 9th - 12th grade, no diploma
    ☐ High school graduate or GED completed
    ☐ Some college credit but no degree
    ☐ Associate degree (e.g., AA, AS)
    ☐ Bachelor's degree (e.g., BA, AB, BS)
    ☐ Master's degree (e.g., MA, MS, MEng, MEd, MSW, MBA)
    ☐ Doctorate (e.g., PhD, EdD) or Professional degree (e.g., MD, DDS, DVM, LLB, JD)

21. MOTHER OF HISPANIC ORIGIN? (Check the box that best describes whether the mother is Spanish/Hispanic/Latina. Check the "No" box if mother is not Spanish/Hispanic/Latina)
    ☐ No, not Spanish/Hispanic/Latina
    ☐ Yes, Mexican, Mexican American, Chicana
    ☐ Yes, Puerto Rican
    ☐ Yes, Cuban
    ☐ Yes, other Spanish/Hispanic/Latina (Specify) _____

22. MOTHER'S RACE (Check one or more races to indicate what the mother considers herself to be)
    ☐ White
    ☐ Black or African American
    ☐ American Indian or Alaska Native (Name of the enrolled or principal tribe) _____
    ☐ Asian Indian
    ☐ Chinese
    ☐ Filipino
    ☐ Japanese
    ☐ Korean
    ☐ Vietnamese
    ☐ Other Asian (Specify) _____
    ☐ Native Hawaiian
    ☐ Guamanian or Chamorro
    ☐ Samoan
    ☐ Other Pacific Islander (Specify) _____
    ☐ Other (Specify) _____

### FATHER
23. FATHER'S EDUCATION (Check the box that best describes the highest degree or level of school completed at the time of delivery)
    ☐ 8th grade or less
    ☐ 9th - 12th grade, no diploma
    ☐ High school graduate or GED completed
    ☐ Some college credit but no degree
    ☐ Associate degree (e.g., AA, AS)
    ☐ Bachelor's degree (e.g., BA, AB, BS)
    ☐ Master's degree (e.g., MA, MS, MEng, MEd, MSW, MBA)
    ☐ Doctorate (e.g., PhD, EdD) or Professional degree (e.g., MD, DDS, DVM, LLB, JD)

24. FATHER OF HISPANIC ORIGIN? (Check the box that best describes whether the father is Spanish/Hispanic/Latino. Check the "No" box if father is not Spanish/Hispanic/Latino)
    ☐ No, not Spanish/Hispanic/Latino
    ☐ Yes, Mexican, Mexican American, Chicano
    ☐ Yes, Puerto Rican
    ☐ Yes, Cuban
    ☐ Yes, other Spanish/Hispanic/Latino (Specify) _____

25. FATHER'S RACE (Check one or more races to indicate what the father considers himself to be)
    ☐ White
    ☐ Black or African American
    ☐ American Indian or Alaska Native (Name of the enrolled or principal tribe) _____
    ☐ Asian Indian
    ☐ Chinese
    ☐ Filipino
    ☐ Japanese
    ☐ Korean
    ☐ Vietnamese
    ☐ Other Asian (Specify) _____
    ☐ Native Hawaiian
    ☐ Guamanian or Chamorro
    ☐ Samoan
    ☐ Other Pacific Islander (Specify) _____
    ☐ Other (Specify) _____

Mother's Name

Mother's Medical Record No.

26. PLACE WHERE BIRTH OCCURRED (Check one)
    ☐ Hospital
    ☐ Freestanding birthing center
    ☐ Home Birth: Planned to deliver at home? ☐ Yes ☐ No
    ☐ Clinic/Doctor's office
    ☐ Other (Specify) _____

27. ATTENDANT'S NAME, TITLE, AND NPI
    NAME: _____ NPI: _____
    TITLE: ☐ MD  ☐ DO  ☐ CNM/CM  ☐ OTHER MIDWIFE
    ☐ OTHER (Specify) _____

28. MOTHER TRANSFERRED FOR MATERNAL MEDICAL OR FETAL INDICATIONS FOR DELIVERY? ☐ Yes ☐ No
    IF YES, ENTER NAME OF FACILITY MOTHER TRANSFERRED FROM:
    _____

REV. 11/2003

# APPENDIX D

## MOTHER

**29a. DATE OF FIRST PRENATAL CARE VISIT**
___/___/___  ☐ No Prenatal Care
MM  DD  YYYY

**29b. DATE OF LAST PRENATAL CARE VISIT**
___/___/___
MM  DD  YYYY

**30. TOTAL NUMBER OF PRENATAL VISITS FOR THIS PREGNANCY**
_____ (If none, enter "0".)

**31. MOTHER'S HEIGHT**
_____ (feet/inches)

**32. MOTHER'S PREPREGNANCY WEIGHT**
_____ (pounds)

**33. MOTHER'S WEIGHT AT DELIVERY**
_____ (pounds)

**34. DID MOTHER GET WIC FOOD FOR HERSELF DURING THIS PREGNANCY?** ☐ Yes ☐ No

**35. NUMBER OF PREVIOUS LIVE BIRTHS** (Do not include this child)

| 35a. Now Living | 35b. Now Dead |
|---|---|
| Number _____ | Number _____ |
| ☐ None | ☐ None |

**36. NUMBER OF OTHER PREGNANCY OUTCOMES** (spontaneous or induced losses or ectopic pregnancies)

36a. Other Outcomes
Number _____
☐ None

**37. CIGARETTE SMOKING BEFORE AND DURING PREGNANCY**
For each time period, enter either the number of cigarettes or the number of packs of cigarettes smoked. IF NONE, ENTER "0".

Average number of cigarettes or packs of cigarettes smoked per day.
# of cigarettes    # of packs
Three Months Before Pregnancy _____ OR _____
First Three Months of Pregnancy _____ OR _____
Second Three Months of Pregnancy _____ OR _____
Third Trimester of Pregnancy _____ OR _____

**38. PRINCIPAL SOURCE OF PAYMENT FOR THIS DELIVERY**
☐ Private Insurance
☐ Medicaid
☐ Self-pay
☐ Other (Specify) _____

**35c. DATE OF LAST LIVE BIRTH**
___/___
MM  YYYY

**36b. DATE OF LAST OTHER PREGNANCY OUTCOME**
___/___
MM  YYYY

**39. DATE LAST NORMAL MENSES BEGAN**
___/___/___
MM  DD  YYYY

**40. MOTHER'S MEDICAL RECORD NUMBER**
_____

## MEDICAL AND HEALTH INFORMATION

**41. RISK FACTORS IN THIS PREGNANCY** (Check all that apply)

Diabetes
☐ Prepregnancy (Diagnosis prior to this pregnancy)
☐ Gestational (Diagnosis in this pregnancy)

Hypertension
☐ Prepregnancy (Chronic)
☐ Gestational (PIH, preeclampsia)
☐ Eclampsia

☐ Previous preterm birth

☐ Other previous poor pregnancy outcome (Includes perinatal death, small-for-gestational age/intrauterine growth restricted birth)

☐ Pregnancy resulted from infertility treatment-If yes, check all that apply:
  ☐ Fertility-enhancing drugs, Artificial insemination or Intrauterine insemination
  ☐ Assisted reproductive technology (e.g., in vitro fertilization (IVF), gamete intrafallopian transfer (GIFT))

☐ Mother had a previous cesarean delivery
If yes, how many _____

☐ None of the above

**42. INFECTIONS PRESENT AND/OR TREATED DURING THIS PREGNANCY** (Check all that apply)
☐ Gonorrhea
☐ Syphilis
☐ Chlamydia
☐ Hepatitis B
☐ Hepatitis C
☐ None of the above

**43. OBSTETRIC PROCEDURES** (Check all that apply)
☐ Cervical cerclage
☐ Tocolysis

External cephalic version:
☐ Successful
☐ Failed

☐ None of the above

**44. ONSET OF LABOR** (Check all that apply)
☐ Premature Rupture of the Membranes (prolonged, ≥12 hrs.)
☐ Precipitous Labor (<3 hrs.)
☐ Prolonged Labor (≥ 20 hrs.)
☐ None of the above

**45. CHARACTERISTICS OF LABOR AND DELIVERY** (Check all that apply)
☐ Induction of labor
☐ Augmentation of labor
☐ Non-vertex presentation
☐ Steroids (glucocorticoids) for fetal lung maturation received by the mother prior to delivery
☐ Antibiotics received by the mother during labor
☐ Clinical chorioamnionitis diagnosed during labor or maternal temperature ≥38°C (100.4°F)
☐ Moderate/heavy meconium staining of the amniotic fluid
☐ Fetal intolerance of labor such that one or more of the following actions was taken: in-utero resuscitative measures, further fetal assessment, or operative delivery
☐ Epidural or spinal anesthesia during labor
☐ None of the above

**46. METHOD OF DELIVERY**

A. Was delivery with forceps attempted but unsuccessful?
☐ Yes ☐ No

B. Was delivery with vacuum extraction attempted but unsuccessful?
☐ Yes ☐ No

C. Fetal presentation at birth
☐ Cephalic
☐ Breech
☐ Other

D. Final route and method of delivery (Check one)
☐ Vaginal/Spontaneous
☐ Vaginal/Forceps
☐ Vaginal/Vacuum
☐ Cesarean
If cesarean, was a trial of labor attempted?
☐ Yes
☐ No

**47. MATERNAL MORBIDITY** (Check all that apply)
(Complications associated with labor and delivery)
☐ Maternal transfusion
☐ Third or fourth degree perineal laceration
☐ Ruptured uterus
☐ Unplanned hysterectomy
☐ Admission to intensive care unit
☐ Unplanned operating room procedure following delivery
☐ None of the above

## NEWBORN INFORMATION

### NEWBORN

**48. NEWBORN MEDICAL RECORD NUMBER**
_____

**49. BIRTHWEIGHT** (grams preferred, specify unit)
_____
☐ grams  ☐ lb/oz

**50. OBSTETRIC ESTIMATE OF GESTATION:**
_____ (completed weeks)

**51. APGAR SCORE:**
Score at 5 minutes: _____
If 5 minute score is less than 6,
Score at 10 minutes: _____

**52. PLURALITY** - Single, Twin, Triplet, etc.
(Specify) _____

**53. IF NOT SINGLE BIRTH** - Born First, Second, Third, etc. (Specify) _____

**54. ABNORMAL CONDITIONS OF THE NEWBORN** (Check all that apply)

☐ Assisted ventilation required immediately following delivery
☐ Assisted ventilation required for more than six hours
☐ NICU admission
☐ Newborn given surfactant replacement therapy
☐ Antibiotics received by the newborn for suspected neonatal sepsis
☐ Seizure or serious neurologic dysfunction
☐ Significant birth injury (skeletal fracture(s), peripheral nerve injury, and/or soft tissue/solid organ hemorrhage which requires intervention)
☐ None of the above

**55. CONGENITAL ANOMALIES OF THE NEWBORN** (Check all that apply)
☐ Anencephaly
☐ Meningomyelocele/Spina bifida
☐ Cyanotic congenital heart disease
☐ Congenital diaphragmatic hernia
☐ Omphalocele
☐ Gastroschisis
☐ Limb reduction defect (excluding congenital amputation and dwarfing syndromes)
☐ Cleft Lip with or without Cleft Palate
☐ Cleft Palate alone
☐ Down Syndrome
  ☐ Karyotype confirmed
  ☐ Karyotype pending
☐ Suspected chromosomal disorder
  ☐ Karyotype confirmed
  ☐ Karyotype pending
☐ Hypospadias
☐ None of the anomalies listed above

**56. WAS INFANT TRANSFERRED WITHIN 24 HOURS OF DELIVERY?** ☐ Yes ☐ No
IF YES, NAME OF FACILITY INFANT TRANSFERRED TO: _____

**57. IS INFANT LIVING AT TIME OF REPORT?**
☐ Yes ☐ No ☐ Infant transferred, status unknown

**58. IS THE INFANT BEING BREASTFED AT DISCHARGE?**
☐ Yes ☐ No

Mother's Name _____
Mother's Medical Record No. _____

# References

Arias, E. (2007). United States life tables, 2004. *National Vital Statistics Reports 54*(14), 1–40.

Banks, J., M. Marmot, Z. Oldfield, and J. P. Smith (2006). Disease and disadvantage in the United States and in England. *Journal of the American Medical Association 295*(17), 2037–2045.

Collins, S. D., W. H. Frost, M. Gover, and E. Sydenstricker (1930). *Mortality from Influenza and Pneumonia in the 50 Largest Cities of the United States* (First ed.). Washington, DC: U.S. Government Printing Office.

Division of Vital Statistics (2000, April). *Report of the Panel to Evaluate the U.S. Standard Certificates.* Addenda, November 2001. Hyattsville, MD: National Center for Health Statistics. Available: http://www.cdc.gov/nchs/data/dvs/panelreport_acc.pdf.

Division of Vital Statistics (2002a). Making vital statistics more vital supplement: The new death certificate—the last word. Summary PowerPoint presentation posted at http://www.cdc.gov/nchs/ppt/dvs/THE NEW DEATH CERTIFICATE.ppt. Document is undated but electronic file carries "last modified" date of August 7, 2002.

Division of Vital Statistics (2002b). The new birth certificate: Making vital statistics more vital. Summary PowerPoint presentation posted at http://www.cdc.gov/nchs/ppt/dvs/THE NEW BIRTH CERTIFICATE.ppt. Document is undated but electronic file carries "last modified" date of September 23, 2002.

Division of Vital Statistics (2004, May 7). *NCHS Procedures for Multiple-Race and Hispanic Origin Data: Collection, Coding, Editing, and Transmitting.* Hyattsville, MD: National Center for Health Statistics. Available: http://www.cdc.gov/nchs/data/dvs/Multiple_race_documentation_5-10-04.pdf.

Division of Vital Statistics (2006). *Guide to Completing the Facility Worksheets for the Certificate of Live Birth and Report of Fetal Death (2003 revision)*. Hyattsville, MD: National Center for Health Statistics.

Friedman, D. J. (2007, June). *Assessing Changes in the Vital Records and Statistics Infrastructure*. Silver Spring, MD: National Association for Public Health Statistics and Information Systems. Available: http://www.naphsis.org/index.asp?bid=984.

Guttmacher Institute (2008, January). Facts on induced abortion in the United States. Available: http://www.guttmacher.org/pubs/fb_induced_abortion.html.

Hamilton, B. E. and S. J. Ventura (2007, May 3). Characteristics of births to single- and multiple-race women: California, Hawaii, Pennsylvania, Utah, and Washington, 2003. *National Vital Statistics Reports* 55(15). Available: http://www.cdc.gov/nchs/data/nvsr/nvsr55/nvsr55_15.pdf.

Hetzel, A. M. (1997). *U.S. Vital Statistics System: Major Activities and Developments, 1950–95*. Includes reprint of "History and Organization of the Vital Statistics System" (1950). HHS Publication No. (PHS) 97-1003. Hyattsville, MD: National Center for Health Statistics.

Hoffman, L. (2004, April). *National Forum on Education Statistics History*. Washington, DC: National Center for Education Statistics. Available: http://nces.ed.gov/forum/pdf/forum_history.pdf [April 2009].

Hummer, R. A., C. B. Nam, and R. G. Rogers (1998). Adult mortality differentials associated with cigarette smoking in the USA. *Population Research and Policy Review* 17(3), 285–304.

Ingram, D. D., J. D. Parker, N. Schenker, J. A. Weed, B. E. Hamilton, E. Arias, and J. H. Madans (2003). United States census population with bridged race categories. *Vital Health Statistics* 2(135). Available: http://www.cdc.gov/nchs/data/series/sr_02/sr02_135.pdf.

Institute of Medicine and National Research Council (2003). *Describing Death in America: What We Need to Know*. June R. Lunney, Kathleen M. Foley, Thomas J. Smith, and Hellen Gelband, eds. National Cancer Policy Board, Institute of Medicine, and Division of Earth and Life Sciences, National Research Council. Washington, DC: The National Academies Press.

Kung, H. C., D. L. Hoyert, J. Q. Xu, and S. L. Murphy (2008, January). Deaths: Final data for 2005. *National Vital Statistics Reports* 56(10). Available: http://www.cdc.gov/nchs/data/nvsr/nvsr56/nvsr56_10.pdf.

Last Acts Partnership (2004). *A Call for a Revitalized National Mortality Followback Survey*. Washington, DC: Last Acts Partnership.

MacDorman, M. F., M. L. Munson, and S. Kirmeyer (2007). Fetal and perinatal mortality, United States, 2004. *National Vital Statistics Reports* 56(3). Available: http://www.cdc.gov/nchs/data/nvsr/nvsr56/nvsr56_03.pdf.

Martin, J. A., B. E. Hamilton, P. D. Sutton, S. Ventura, F. Menacker, S. Kirmeyer, and M. L. Munson (2007, December). Births: Final data for 2005. *National Vital Statistics Reports 56*(6). Available: http://www.cdc.gov/nchs/data/nvsr/nvsr56/nvsr56_06.pdf.

Menacker, F. and J. A. Martin (2008, February). Expanded health data from the new birth certificate, 2005. *National Vital Statistics Reports 56*(13). Available: http://www.cdc.gov/nchs/data/nvsr/nvsr56/nvsr56_13.pdf.

Mills, C. E., J. M. Robins, and M. Lipsitch (2004, December 16). Transmissibility of 1918 pandemic influenza. *Nature 16*, 904–906.

Mulder, T. J. (2002, July). *Accuracy of the U.S. Census Bureau National Population Projections and Their Respective Components of Change.* Working Paper Series No. 50, published online at http://www.census.gov/population/www/documentation/twps0050/twps0050.html. Washington, DC: U.S. Census Bureau.

National Commission on Terrorist Attacks Upon the United States (2004). *The 9/11 Commission Report.* Final Report. Washington, DC: U.S. Government Printing Office.

Paulson, J., W. Ramsini, E. Conrey, R. Duffy, and M. P. Cooper (2007, October 26). Unregistered deaths among extremely low birthweight infants—Ohio, 2006. *MMWR 56*(42), 1101–1103.

Retherford, R. D. (1972). Tobacco smoking and the sex mortality differential. *Demography 9*(2), 203–216.

Rogers, R. G., R. A. Hummer, and C. B. Nam (2000). *Living and Dying in the USA: Behavioral, Health, and Social Differentials of Adult Mortality.* New York: Academic Press.

Rosenberg, H. M. (2008). Selected successes of the Vital Statistics Cooperative Program. Paper to accompany presentation at Workshop on Vital Data for National Needs; published at http://www.naphsis.org/NAPHSIS/files/ccLibraryFiles/Filename/000000000678/SELECTED%20SUCCESSES%20OF%20THE%20VITAL%20STATISTICS%20COOPERATIVE%20PROGRAM.pdf [March 2009].

Rothwell, C. J. (2004, October). Reengineering vital registration and statistics systems for the United States. *Preventing Chronic Disease 1*(4). Available: http://www.cdc.gov/pcd/issues/2004/oct/pdf/04_0074.pdf.

Sutton, P. D. (2008). Births, marriages, divorces, and deaths: Provisional data for September 2007. *National Vital Statistics Reports 56*(18). Available: http://www.cdc.gov/nchs/data/nvsr/nvsr56/nvsr56_18.htm.

Tolson, G. C., J. M. Barnes, G. A. Gay, and J. L. Kowaleski (1991). The 1989 revision of the U.S. standard certificates and reports. *Vital and Health Statistics 4*(28). Available: http://www.cdc.gov/nchs/data/series/sr_04/sr04_028.pdf.

Townsend, P. (1987). Deprivation. *Journal of Social Policy* 16(2), 125–146.

Townsend, P., P. Phillimore, and A. Beattie (1988). *Health and Deprivation: Inequality and the North*. London: Croom Helm.

U.S. Department of Health and Human Services (2000, November). *Healthy People 2010* (2nd ed.). With Understanding and Improving Health and Objectives for Improving Health. 2 vols. Washington, DC: U.S. Government Printing Office.

U.S. Office of Management and Budget (1997, October 30). Revisions to the standards for the classification of federal data on race and ethnicity. *Federal Register 62*, 58781–58790. Revision of Statistical Policy Directive No. 15.

## COMMITTEE ON NATIONAL STATISTICS

The Committee on National Statistics was established in 1972 at the National Academies to improve the statistical methods and information on which public policy decisions are based. The committee carries out studies, workshops, and other activities to foster better measures and fuller understanding of the economy, the environment, public health, crime, education, immigration, poverty, welfare, and other public policy issues. It also evaluates ongoing statistical programs and tracks the statistical policy and coordinating activities of the federal government, serving a unique role at the intersection of statistics and public policy. The committee's work is supported by a consortium of federal agencies through a National Science Foundation grant.